ENERGY OF EXISTENCE

AUDREY E. RANDLES

Copyright © 2012 Audrey Elizabeth Randles

All rights reserved.

ISBN: 9798666114117

Introduction

The theory of Matrix is a cosmological theory considering a new understanding of space and time, along with the Space-Time and energy structure of the existing objects and systems.

The space of the Universe is filled with a multiplicity of space-objects. The time of the Universe is filled with a multitude of time-objects. Objects and systems, mainly developing time and accumulating tremendous potential energy resources, are still invisible in the dark.

The Theory of Matrix series of books offers the exiting developments in cosmological theory.

We combine elements of psychology, cosmology, and astrophysics to discover secrets hidden deep in the Universe.

'Energy of Existence' is the 3rd book of the series.

In this book, we discuss the energy structure of the multimodal objects and systems and relations between energy and matter, space and time, and gravity and antigravity.

If you keep an open mind, this book is for you.

Stay well, and enjoy your reading and your future discoveries.

Yours sincerely,
Audrey Elizabeth Randles
JULY 29, 2020

*Image 1: 'Dust in the Quasar Wind (Artist Concept)'
Image credit: NASA/JPL-Caltech*

'Using Spitzer's infrared spectrograph instrument, scientists found a wealth of dust grains in a quasar called PG2112+059 located at the center of a galaxy 8 billion light-years away. The grains - including corundum (sapphires and rubies); forsterite (peridot); and periclase (naturally occurring in marble) - are not typically found in galaxies without quasars, suggesting they might have been freshly formed in the quasar's winds.' NASA

Acknowledgement

We would like to express our gratitude to the National Aeronautics and Space Administration (NASA), NASA's Jet Propulsion Laboratory (JPL), the University of California at Los Angeles (UCLA), the 2MASS mission, and the California Institute of Technology (Caltech) for the impressive space image and exciting descriptions. We would like to express our gratitude to the Oxford Dictionary for the precise formulation of the tendencies existing in the modern natural sciences.

The views and opinions of the author, expressed in this book, do not necessarily state or reflect those of the Oxford Dictionary, NASA's Jet Propulsion Laboratory, the University of California at Los Angeles, the 2MASS mission, the California Institute of Technology or National Aeronautics and Space Administration.

Image 2: 'NEOWISE: Back to Hunt More Asteroids (Artist Concept)' Image Credit: NASA/JPL-Caltech

Contents

Introduction
Acknowledgement
Chapter 1 Energy and Information
Chapter 2 Relativity of Mass and Info-Energy
Chapter 3 Zero-point Energy
Chapter 4 Latent Info-Energy
Chapter 5 Actual Info-Energy
Chapter 6 Energy Curves Space-Time
Chapter 7 Matrix Balance
Chapter 8 Time Dimensions
Chapter 9 Matrix Imbalance
Chapter 10 Matrix Forces
Chapter 11 Time and Energy Flow
Chapter 12 Balancing Transformations
Chapter 13 The Theory Of Light
Chapter 14 The Genome of the Universe
Chapter 15 Dark Matter And Dark Energy
Chapter 16 Black Matter and Black Energy
Chapter 17 Reverse of the Time and Energy Flow
Chapter 18 Rotation
Chapter 19 Main Principles
Chapter 20 Natural Laws of the Theory of Matrix
Afterword
Content Use Policy

Energy and Information

Different models of the early Universe, Big Bang Theory, and theories involving singularities do not explain where the matter did come from to become the uniform mass or any other dense medium. They don't explain where the initial density perturbations did come from to grow under their gravity. The Big Bang Theory has two weak points - the beginning of the Universe existence from nothing and the end of the Universe existence to nothing. Do you remember Émilie du Châtelet? The total energy is conserved over time. The Big Bang Theory violates the law of conservation of energy. Energy cannot be created or destroyed. Energy is being transformed from one form to another and transferred. The theory of singularity is left with the same problem - the start of the singularity. The theories of singularity and Big Bang reflect limitations of modern mathematics.

Nothing is coming from nothing and going to nothing. Energy and information cannot be lost or appear from a singularity. There is no need for the Universe beginning from a Big Bang and Universe final collapse. Space-Time is complete and harmonic, and characteristics of the Universe are different in different modalities of Space-Time.

The existence of objects and systems rises in the Space-Time, Info-Energy, and Mass-Energy imbalance.

Energy and information are the fundamental characteristics of the objects and systems existing in the Universe. Space-Time is built by and filled with energy. We could cancel time and space altogether and represent the world as energy - then, accordingly, all dynamics, forms, or limits, masses and volumes should be cancelled. We would have a world of homogeneous energy without any other quality, so the meaning of the world and energy would cease to exist.

Energy is a quality of information, and information is a quality of energy. Information is data that is always associated, processed, stored or transmitted along with energy or matter.

Energy and information cannot be lost. Energy exists in the latent, or potential, and actual forms set by information. Information also takes different forms and exists in the potential and actual forms fixed by energy.

We use the term 'Info-Energy' (forms of energy) to describe the energy and information, which are complementary and cannot be separated. The density of information is proportional to the density of energy associated with the information.

Mass-Energy is 'the mass of a body regarded as energy, according to the laws of relativity' (Oxford Dictionary). We, respecting the laws of relativity, use the term 'Mass-Energy' as interchangeable with the term 'Info-Energy' of an object or a system.

Energy and information are the fundamental characteristics of the Universe and the existing objects and systems. Space-Time is built by and filled with energy. Please see my book 'Space and Time' for the details associated with Space-Time in the Universe.

Space and mass are associated with actuality. We can directly perceive energy acting in space. For example, you can directly perceive the volume and weight of the book you keep in your hand. Space, or the volumes of objects and systems, is filled with actual Info-Energy. The actual Info-Energy of an object is an integrated form of energy represented in mass, kinetic energy, and other energy representations, actually acting in the volume of the object in the period of time now occurring.

Time and time-related Info-Energy are associated with latency and potentiality. We cannot directly perceive time-related latent, or potential, energy and associated information. For example, we cannot right now taste the dinner that we had a week ago. On the other hand, we cannot, at this point, read a book we will first glance at tomorrow. Both the dinner and the book are latent to us - we cannot touch or see them directly at the moment. Time and time-related Info-Energy are latent to us. Some objects and systems, events and their changes lie in the past. Other events, objects and systems are waiting for us in the future.

Time and time-related Info-Energy are associated with latency or potentiality as a capacity to perform work. When the object or the system's capacity to perform work is being realized - energy is associated with space. Similarly, water, filling up an empty container, takes shape.

Energy, existing in space, takes perceivable forms creating sensible information. We sense different forms of energy as masses, kinetic energy, and other energy representations acting in our dynamic world in the period of time now occurring. The potential and actual Info-Energy structures counteract and keep a balance in the multimodal structures of the objects and systems, retaining their total Space-Time and Info-Energy unchanged.

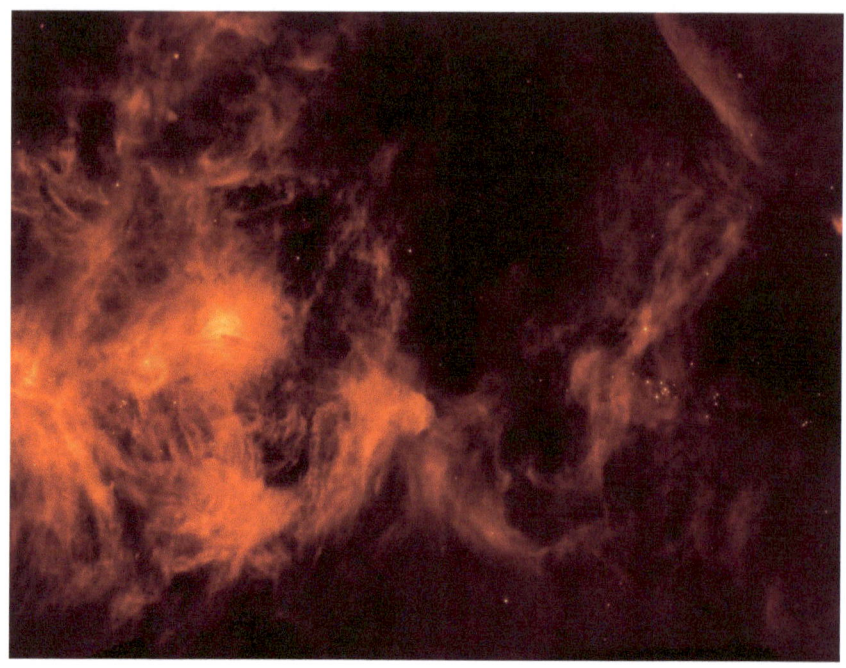

Image 3: 'Perseus Molecular Cloud' Image credit: NASA/ JPL-Caltech

'A collection of gas and dust over 500 light-years across, the Perseus Molecular Cloud hosts an abundance of young stars. It was imaged here by the NASA's Spitzer Space Telescope... Located on the edge of the Perseus Constellation, the collection of gas and dust is about 1,000 light-years from Earth and about 500 light-years wide.' NASA

Relativity of Mass and Info-Energy

Time, space, mass, energy, and information do not exist independent of objects and systems of the objects. They are the properties of the objects and systems. Time, space, mass, energy, and information are to be defined in relation to a frame of reference.

The concept of relativity and Einstein's Special Theory of Relativity state that all motion is relative and that the velocity of light in a vacuum has a constant value which nothing can exceed.

We add that time is relative to space, changing, and transforming into space. And space is relative to time, changing, and transforming into time. The universal proportionality, but not equality, exists between corresponding amounts of time and space. The volume of a system may change, although the total Space-Time of the system remains constant. Time of a system may change, although the total Space-Time of the system remains constant. Please see my book 'Space and Time' for details related to Space-Time Relativity.

The latent, or potential, energy is relative to the actual energy, changing, and transforming into the actual energy; and the actual energy is relative to the latent energy, changing, and transforming into the latent energy. For example, in the case of the Earth gravity, the potential energy is the energy possessed by a body under its position relative to the Earth. Being in motion, or in the falling to the Earth, the body possesses kinetic energy.

The system's information represents the forms of energy 'in-formation' in our dynamic world. The latent, or potential, information is relative to the actual information, changing, and transforming into the actual information; and the actual information is relative to the potential information, changing, and transforming into the potential information. Transformations of information proceed along with energy transformations.

Overall, potentiality is relative to actuality, changing and transforming into actuality - the reality of our dynamic world. Moreover, actuality is relative to potentiality, changing and transforming into potentiality or latency.

Albert Einstein has specified the connection between mass and energy of the object, 'If an amount of energy E be given to a body, the inertial mass of the body increases by an amount E/c^2, where c is the velocity of light in vacuo. On the other hand, a body of mass m is to be regarded as a store of energy of magnitude mc^2.' [Einstein Albert, A Brief Outline of the Development of the Theory of Relativity (1921)]

The universal proportionality exists between equivalent amounts of energy and mass,

$E = mc^2$,

where E is energy and m is mass.

Energy is relative to mass, changing, and transforming into a mass, and mass is relative to energy, changing, and transforming into energy.

We add that reading the famous equation for mass-energy equivalence $E = mc^2$ with respect to the 2-dimensional Space-Time helps consider the Theory of Matrix. Energy is associated with the 2-dimensional time the same way as mass, corrected with the squared numerical value of the speed of electromagnetic waves propagation in a vacuum, is associated with the 2-dimensional space,

$Et^2 = [c]^2 ml^2$,

where E is energy, m is mass, l^2 is the 2-dimensional space, t^2 is the 2-dimensional time, and $[c]^2$ is the Coefficient of Transformation.

Rest mass and rest energy remain proportional to one another the same way as the 2-dimensional space and the 2-dimensional time remain proportional to one another.

The total energy is conserved over time. According to the law of conservation of energy, energy cannot be created or destroyed. Energy is being transformed from one form to another and transferred. Transformations of the information, depicting the form of energy or mass, proceed along with energy transformations.

Image 4: 'The Milky Way Centre Aglow with Dust' Image credit: NASA/JPL-Caltech

Zero-point energy

Energy does not exist independent of space and time. We mentioned in the Introduction that Space-Time is built by and filled with energy. We could cancel time and space altogether and represent the world as energy - then, accordingly, all dynamics, forms, or limits, masses and volumes should be cancelled. We would have a world of homogeneous energy without any other quality, so the meaning of the world and energy would cease to exist.

Theoretically, a '0' time-point, or 'a time-null-point', permits maximum space of the system. A '0' potential energy point is associated with the '0' time-point. Accordingly, the '0' potential energy point permits maximal actual energy that fills the volume of the system. For example, all the Universe exists at this time-point - this moment. It exists 'now', at the same moment for all the Universe with all its space, masses, stars, and galaxies.

Furthermore, a '0' space-point, or 'a space-null-point' - a point 'here', permits maximum time. A '0' actual energy point is associated with the '0' space-point. Accordingly, the '0' actual energy point permits maximal potential energy of the system. For example, all the time of the Universe exists 'here' - at this point in space. It exists at the end of the needle - the same point in space for all potentiality of time with all its capacity to perform work, with all the past and unlimited opportunities of the current time and the future. We could build the Matrix with a space-null-point - the point 'here' and the associated time volume.

The coexisting 'o' space-points and 'o' time-points, along with 'o' actual energy points and 'o' potential energy points, cancel out one another along with the start and the end of space, time, and energy. It opens the unlimited actuality and potentiality, nullified in the ultimate balance of the Grand Universe.

We think about the Universe as the infinite space, existing in various combinations at this moment and every other moment in time. We cannot imagine the Universe as the infinite time that exists in different arrangements at this point and every other point in space.

'By laying down the relativity postulate from the outset, sufficient means have been created for deducing henceforth the complete series of laws of Mechanics from the principle of conservation of energy (and statements concerning the form of the energy) alone.' [Minkowski Hermann, The Fundamental Equations for Electromagnetic Processes in Moving Bodies. Appendix (1908)]

We must not put the principle of conservation of energy aside and ignore the basics of the quantum theory.

The 'o' Space-Time and 'o' energy point, along with the associated 'o' information, as the start or the end of the Universe, will never be found or proven because they never exist. We can accept them as the point of the balance of the multimodal Grand Universe, but they are not achievable in our thermodynamic Universe as we sense and measure it. The ideas of zero-point energy are related exclusively to the human perception of time as a point 'now'.

The speed of light in vacuum and the Planck units, based on the calculations using the speed of light in vacuum as the fundamentals proven by experiments - Planck length, Planck time, and Planck energy, provide the upper and lower limits for the existence of the Universe as we measure and sense it.

The speed of light in vacuum sets the fundamental relation between 1-dimensional space and 1-dimensional time that defines the exact proportion of space to time as the upper limit of our dynamic world. This fundamental relation is supported by the energy divergence between space-associated actual energy and time-associated potential energy.

On the other hand, Planck length, Planck time, and Planck energy reflect the minimal Space-Time and Info-Energy imbalance in the existing objects and systems as the necessary deviations from the '0' Space-Time point and '0' energy point.

Planck length, Planck time, and Planck energy, based on the calculations using the speed of light in vacuum as the fundamentals proven by experiments, reflect the minimal Space-Time and Info-Energy imbalance in the objects and systems existing in our dynamic world. These Planck units define the exact lower limits as the conditions of an object or a system's existence and, accordingly, factors supporting the reverse of the Space-Time and Info-Energy flow in the thermodynamic Universe and the existing objects and systems as we measure and sense them. These limits influence the direction of the Space-Time and Info-Energy flow in the Universe and the existing objects and systems. We will talk about it in details in the following chapters.

Latent Info-Energy

There is no empty space in the Universe. The Universe is filled with energy. We perceive this energy in different forms, such as mass, kinetic energy, and other energy representations, acting in the volume of the Universe in the period of time now occurring. The term 'latent energy' reflects the current human understanding of energy as 'the property of matter and radiation, which manifests a capacity to perform work' (Oxford Dictionary).

The latent, or potential, energy of the Universe is associated with time. It represents the Universe' capacity to perform work. The Universe' latent, or potential, information, coding and decoding a form of energy, is fixed by the Universe' latent energy within its time and potential space. The latent, or potential, Info-Energy is the predominant form of energy in our Universe.

Latent energy, taking a form at the point of divergence between space and time, draws space and time. Space and time are coming from the background and become visible like a light-sensitive material.

Latent Info-Energy configures the potential space and time of the Universe and the existing objects and systems, including stars and planets, systems holding Black Holes, galaxies and star clusters, atoms, electrons, and other objects and systems existing in the Universe.

We add that latent energy manifests the particular capacity (to perform work) that has been existing in the past, is existing 'now', at this time-point, and the capacity to perform work in the future.

Accordingly, the latent form of energy is associated with the potential space and time of an existing object (or a system), including its past, present, and future.

If the object exists now, we can measure its properties in the period of time now occurring. Still, the object's past characteristics, its future alterations, and the object's likely current changes are inaccessible to the direct measurement we undertake at the current moment. Consider the following examples of the associations between the volume of the system filled with energy and the potential space underlying the volume of the system.

Do you still have petrol that was used on the way home yesterday? Petrol that has been used on the way home yesterday does not exist in actuality at the current moment.

Didn't you use petrol on the way home yesterday? The petrol that was not used on the way home yesterday and the petrol, used on the way home yesterday, exist as a possibility, potentiality, or the time-associated latency that influences the past, present, and future. It might be that petrol was not used yesterday, and so you are getting your car back on the road now.

How many containers can be filled at a petrol station? The containers which can be filled at the petrol station, and those containers which cannot be filled, along with containers, which you will buy and those you will not buy tomorrow, all exist as the potentiality, or the time-related latency that influences the past, present, and future.

The petrol-container system is potentiality or latency that is associated with time and the potential space underlying the volume of the containers that could be filled and could not be filled during the time of their existence. The petrol-container system exists along with a volume of your containers actually occupied and not occupied by petrol now - in the period of time now occurring.

Time of an object (or a system) is built by the object's latent energy that is coded and fixed by the latent information, which the object possesses in time, including its past, present, and future. The time-associated imperceptible Info-Energy forms a timeline of an object or a system's existence.

We, usually, think of any potential event in terms of probability, opportunity or possibility. We cannot directly perceive time and associated Info-Energy. We cannot directly perceive the events, which are in the past or the future. We understand the events of the past and the future as immobile, unchanging, and associated with the timeline. We think nobody can predict the future or change the past. We keep our memories, but we cannot perceive directly any progression or dynamics within the past and the future.

Prof. Hermann Weyl described the Matrix of Light - the Light Cone of the Special Relativity as follows,

'According to the old understanding, the spoken word 'now' intersects not only the course of my inner life into past and future, but it brings this cut by a single stroke into the entire world: it intersects the world in a similar way into two parts, that are without space between each other: the past and the future, like a horizontal plane bisects the space into a lower and upper part.

'According to the relativity principle, however, the bisection of past-future is of a different kind when it is seen from world point o, and it corresponds to the one that in three-dimensional space is caused by a complete circular cone (it is sketched in the vertical projection in the figure; the curved line is the world line of my body, that is of course bisected through 0 into two parts, the past part and the future part of my life).' [Einstein Albert, Lenard Philipp, Weyl Hermann, Gehrcke Ernst, The Bad Nauheim Debate. The Discussion concerning the theory of relativity at the Meeting of Natural Scientists. (1920)]

The objects and systems' time and time-associated latent Info-Energy are represented in their multimodal Space-Time and Info-Energy structures, which we call the object or the system's Matrix for short. The 1-dimensional, 2-dimensional, and 3-dimensional structures of an object or a system are integrated into the object or the system's multimodal Space-Time and Info-Energy structure.

The object's potential Info-Energy is arranged by the 2-dimensional representations of background radiation and light into the object's 2-dimensional Info-Energy grid forming the object's time and potential space. Any object, which is smaller than the wavelength supporting the Matrix grid of the object, does not exist in Space-Time.

Time-associated latent Info-Energy of the 2-dimensional grid is a property of every existing object and every system. It contains the object' time-associated latent energy, which is coded and fixed by the time-associated latent information. The object's time-associated latent Info-Energy forms the object's Spaces of Time, including the past, present, and future. Time and time-related Info-Energy are latent to us. Some objects and systems, events and their changes lie in the past or the future, and we cannot directly perceive time-related potential energy and associated information. For example, we cannot, at this point, read a book we will first glance at tomorrow. Spaces of the past and the future look immobile, unchanging, inaccessible, appear empty and connected with the undifferentiated Continuum. The 1-

dimensional Space-Time and associated Info-Energy of the Continuum fill spaces of the past and the future with the only quality of the duration. The 1-dimensional Space-Time and associated Info-Energy are to be defined in relation to a frame of reference, following a timeline of an object or a system's existence.

The space-associated latent Info-Energy of the 2-dimensional grid forms the object's potential space of the period of time now occurring. The 2-dimensional grid contains the Info-Energy of the potential space that underlies the object's volume.

The potential space of an object (or a system) is a kind of transmitter between the object's time-associated latent Info-Energy and the actual Info-Energy acting in the object's volume. There is neither progression nor emission but a divergence within the object's 2-dimensional structure. The balancing flux of radiation diverges within the object's 2-dimensional Space-Time across the 2-dimensional Info-Energy grid.

The object's potential space acts similar to a membrane of a living cell and structural genes. It is a boundary that is keeping and transforming energy and associated information in compliance with a stored data to balance the object's existing capacity to perform work against its actual realisation in the forms of actual energy represented in mass, kinetic energy, and other energy representations acting in the object's volume in the period of time now occurring.

The 2-dimensional Info-Energy grid of the Universe, accumulating data that is related to the Universe, we call 'the genome of the Universe'. We compare it with a human genome that is the complete set of genetic information for a newborn child. The genome of the Universe keeps the past and future secrets of the body of our young Space-Rising Universe. The thermodynamic Universe is expanding in space. It is 'young' in its energy scale. Latent Info-Energy represents a dominant form of energy existing in the Universe.

The total latent, or potential, Info-Energy of an object or a system is the total latent energy and associated latent information, which the existing object or the system possesses in space and time, including its past, present, and future.

The total latent, or potential, Info-Energy as a particular capacity (to perform work) of an object is integrated, as a part of a coherent whole, into the total latent, or potential, Info-Energy of the system. The total latent, or potential, Info-Energy of the existing objects and systems is integrated into the latent Info-Energy of the Universe.

The object's capacity to perform work, being entirely negligible relative to the system's capacity, is reflected in the system's characteristics. For example, the Earth capacity to perform work in the past, present, and future is entirely negligible relative to the capacity of the Universe.

The latent Info-Energy of the multiple Universe modalities, such as the thermodynamic Universe, Multiverse, other Universes, stages of the Universe development, and other Universe modalities, are finally integrated into the latent Info-Energy of the Grand Universe. Our dynamic world is a limited modality of the Grand Universe.

Actual Info-Energy

Time and time-related Info-Energy are associated with potentiality as a capacity to perform work. Space and mass are associated with the actuality. We sense directly different forms of energy as masses, kinetic energy, and other energy representations acting in our dynamic world in the period of time now occurring and we can measure them. You can directly perceive the volume and weight of the book you keep in your hand now.

When the object or the system's capacity to perform work is being realized - energy is coming from the background. Similarly, water, filling up an empty container, takes shape. Energy, existing in space, takes perceptible forms creating sensible information while acting in the volumes of objects and systems in the micro and macro world as masses, kinetic energy, and other forms of energy.

The actual Info-Energy is a specific property of an object or a system of the objects. We perceive this energy in different forms of energy and matter, which we can measure directly in the period of time now occurring.

The term 'actual Info-Energy' includes all forms of energy and matter, such as mass, kinetic energy, dark energy and dark matter, and other energy representations, acting in the volumes of the existing objects and systems in the period of time now occurring.

The object's total actual energy is an integrated form of energy acting in the volume of the object at a point 'now'. The object's complete actual information, representing the forms of energy, is fixed by the total actual energy in the volume of the object. The density of energy is proportional to the density of the related information.

The object's invariant mass is integrated, as a part of a coherent whole, into the invariant mass of the system and represented in the system's characteristics. The object's total actual Info-Energy is integrated, as a part of a coherent whole, into the actual Info-Energy of the system and represented in the system's characteristics. For example, the mass of your body cell is integrated into the total mass of your physical body. It is entirely negligible relative to the mass of your body. The mass of a water-drop in the ocean is entirely negligible relative to the mass of the Solar system. The mass of the Milky Way is entirely negligible relative to the mass of the Universe.

The total actual Info-Energy of the existing objects and systems, including visually empty spaces, subatomic particles and holes, quanta, stars, planets, galaxies, systems holding Black Holes, other objects and systems existing in the Universe, is integrated, as a part of a coherent whole, into the actual Info-Energy of the Universe.

The volume of the Universe is filled with actual energy and associated information. The Universe actual Info-Energy is represented in mass, kinetic energy, and other energy representations, actually acting in the volume of the Universe in the period of time now occurring.

The total actual Info-Energy of the multiple Universe modalities, such as the thermodynamic Universe, Multiverse, other Universes, stages of the Universe development, and other existing modalities of the Universe, are finally integrated into the total actual Info-Energy of the Grand Universe.

Image 5: 'The 'Serpent' Star-Forming Cloud Spawns Stars' Image credit: NASA/JPL-Caltech/2MASS

'Within the swaddling dust of the Serpens Cloud Core, astronomers are studying one of the youngest collections of stars ever seen in our galaxy. This infrared image combines data from NASA's Spitzer Space Telescope with shorter-wavelength observations from the Two Micron All Sky Survey (2MASS), letting us peer into the clouds of dust wrapped around this stellar nursery.

'At a distance of around 750 light-years, these young stars reside within the confines of the constellation Serpens, or the "Serpent." This collection contains stars of only relatively low to moderate mass, lacking any of the massive and incredibly bright stars found in larger star-forming regions like the Orion nebula.

'The stellar 'hatchlings' in the Serpens Cloud Core represent the very youngest stages of stellar development. They appear as red, orange and yellow points clustered near the center of the image. Other red features include jets of material ejected from these young stars. Some mature stars that are not in the nebula appear yellowish due to dust obscuring our view at shorter, bluer wavelengths.

'This region also includes a population of prenatal stars that are so deeply enshrouded in their dusty cocoons to be completely hidden in this view. They only become detectable at much longer wavelengths of light.' *NASA*

Energy Curves Space-Time

According to the theory of General Relativity, energy curves Space-Time. Please see Einstein Albert, 'Relativity: The Special and General Theory' (1916). We support this idea and provide descriptions of the main mechanisms as precise as possible for a person functioning in the dynamic 3-dimensional world.

We mentioned above that the Matrix is shaped by the regular repeated (with the equal distances, typically rectangular) grid-like 2-dimensional arrangement of the info-energetic net-structure. The infinitely thin filament forms the Matrix of every existing object and every system.

The 2-dimensional Info-Energy grid contains the object/system's potential Info-Energy. The total Info-Energy of the Matrix grid is the total potential energy and associated potential information, which the existing object or the system possesses in space and time, including its past, present, and future. Background radiation, including those known as cosmic background radiation, and light arrange the Info-Energy of the Matrix grid of the existing objects and systems.

The grid composes the fundamental 2-dimensional Space-Time of the Matrix. The potential space of the Current Time forms the 2-dimensional 'container' for an object or a system's volume filled with actual Info-Energy. The Matrix grid limits the object's volume at the centre of the Matrix and shapes the object's Spaces of Time. The Matrix grid has a form of the Riemannian Manifold with the positive or negative curvature, and Riemannian geometry applies to its investigation.

Strictly speaking, the reproduction of the Matrix 2-dimensional grid on the paper is not exact. It is related to the difficulties to reproduce the 2-dimensional structure on the paper.

It is a theoretical mistake that one can currently reproduce a 2-dimensional structure on the paper. A drawing on the paper and PC screen does not reflect a 2-dimensional structure - our world-perception influences on making drawings and PC programming. The relative difference between our perception of space and time makes this world of perceptible forms being sensible to us. Accordingly, the structures, which are reproduced on the paper or PC screen, regardless of how special they would be, are no more than our projections of the 3-dimensional objects. Their proportions and structure do not reflect characteristics of the 2-dimensional structures. Similarly, proportions and directions of the 2-dimensional structure in the multimodal Space-Time are entirely different from our projection of any structure on the surface.

Besides, one cannot create a 2-dimensional structure using 3-dimensional particles, one layer of 3-dimensional cells, crystals or atoms, even with the help of nanotechnology. The closest object to the understanding of the 2-dimensional space structure is an ideal square wave.

The exact reproduction of the Matrixes (Figures 1-11) without damaging the main idea is impossible on the paper or in the computer simulation. Although the shapes of the identified types of Matrixes are correct, we rely entirely on the description.

Two types of the Matrixes have been identified - the Time-Rising Matrix (TRM) and the Space-Rising Matrix (SRM).

The TRMs are the property of the radiating objects and systems, including 'black body' radiation spectrum. The 2-dimensional grid of the TRM has a form of the Riemannian Manifold with the negative curvature (Figure 1).

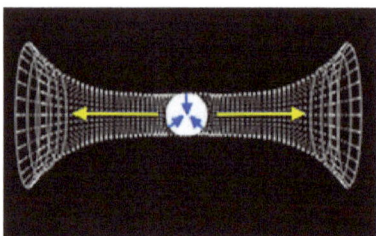

Figure 1: Time-Rising Matrix (TRM)

SRMs are the property of non-radiating objects and systems. The 2-dimensional grid of the SRM has a form of the Riemannian Manifold with the positive curvature (Figure 2).

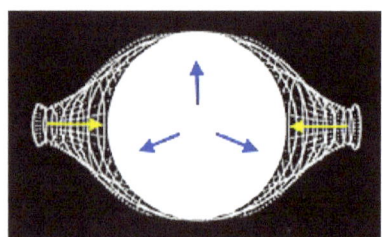

Figure 2: Space-Rising Matrix (SRM)

The Arrows of Time are drawn as the yellow arrows on the reproduction of the Matrix, while the direction of the Space of the Current Time is represented by the blue arrows, situated following the human perception of the objects and systems as exhibiting three Space dimensions (x, y, and z, or a combination of three directions, which can be chosen from the terms: length, width, height, depth, and breadth).

The Space of the Progressive Time is the space of the Matrix cone related to the future of the object (Figures 1-11 - the left cone). The Space of the Regressive Time is the space of the Matrix cone associated with the past of the object (Figures 1-11 - the right cone).

According to Albert Einstein, 'the ponderable masses will be the determining factor in producing the field, or, according to the fundamental result of the Special Theory of Relativity, the energy density...' [Albert Einstein, A Brief Outline of the Development of the Theory of Relativity (1921)] Under the Theory of Matrix, the Matrix is not the field but the structure. The ponderable masses are the determining factors in producing the energy density, information density, and multimodal Space-Time and Info-Energy structures of the existing objects and systems.

The space-time-energy Continuum fills up spaces of our past and the future. The 1-dimensional space-time-energy is a property of every existing object and every system.

The object's Space of the Current Time at the Matrix centre is filled with the actual Info-Energy. We can directly perceive and measure the objects and systems' volumes and the forms of actual energy, such as masses, kinetic energy, and other forms of energy acting in the volumes of objects and systems in our dynamic world in the period of time now occurring. The complete actual information is coded and fixed by the object's actual energy and represented in forms of energy and matter in the volume of the object. The density of information is proportional to the density of energy and matter associated with the information.

The Space of the Current Time equals the volume of the object embedded at the centre of the Matrix. The volume of the object is filled with the total actual Info-Energy, acting in the volume of the object in the period of time now occurring. The complete actual information is coded and fixed by the total actual energy in the volume of the object.

The object's total actual Info-Energy is an integrated form of energy, actually acting in the volume of the object in the period of time now occurring and represented in mass, kinetic energy and other energy representations.

The space-associated latent 2-dimensional Info-Energy of the object's grid forms the potential space of the period of time now occurring. The object's potential space underlies the object's volume within the multimodal Space-Time and Info-Energy structure.

The time-associated 2-dimensional Info-Energy of the object's grid forms the time-limits of the object's existence. It forms the Spaces of Time. Energy, building time, is latent Info-Energy associated with the past, present, and future of the object.

The object's Matrix grid is the 2-dimensional framework of the object. It holds a complete set of data related to the enclosed object, similar to a chromosome of a living cell. It carries information in the form of Info-Energy blocks.

The notion of the curvature is to be defined in a way that is intrinsic to the manifold. The curvature of the 2-dimensional grid depends on the Space-Time, Info-Energy, and Mass-Energy properties of the related object. The curvature of the Matrix does not depend on how the surface is enclosed in 1-dimensional, 3-dimensional or higher-dimensional Space-Time. The curvature of the Matrix does not depend on how the object or the system's volume is inserted at the Matrix centre.

Black pixels, reading zero and corresponding to the latent information, are fixed by the latent energy of the 2-dimensional grid. They are associated with the timing mechanisms and sweep rates, and the address of a pixel corresponds to its Space-Time coordinates.

We suppose that the blocks of the data, involved in the actuality or located relatively close to the point 'now' in time, are situated closer to the centre of the Matrix, than other latent information. Hypothetically, if the blocks of data are associated with the Matrix Space of the Current Time, then they might have a 3-dimensional structure. The distances along the paths and angles are to be measured as the characteristics of the object's latent, or potential, Info-Energy.

The latent energy of the 2-dimensional grid is arranged by the 2-dimensional representations of background radiation, including those currently known as cosmic background radiation, being dynamic in our dynamic world. Accordingly, the Matrix 2-dimensional grid, being the external framework of the object/system in the multimodal Space-Time, is simultaneously the dynamic internal framework and Space-Time and Info-Energy skeleton of this object/system in our dynamic world. An example of the Matrix grid influence on the macro-scale is the subatomic and atomic processes associated with the regulation of heat in the body of the Universe, and the example of the grid-forming energy is the energy structure built by the cosmic microwave background radiation carrying this heat and information associated with the regulation of heat in the Universe and objects and systems existing in the Universe. Consequently, every existing object and every system of the objects is a multimodal structure carrying the properties of different Space-Time modalities.

The Matrixes depict the total Space-Time and Info-Energy of the existing multimodal objects and systems. The total Info-Energy of the Matrix is the total energy and associated information that the existing object/system possesses in space and time, including its past, present and future.

The objects and systems of the objects of limited mass and energy are limited in space and time. Accordingly, the Matrixes for the objects and systems of the objects of limited mass and energy are limited in space and time.

The potential and actual Info-Energy structures counteract and keep a balance at the central point of the Matrix symmetry, and through the Matrix axis under the laws of Space-Time, Info-Energy, and Mass-Energy Conservation, Transformations, Reversibility, Limitation, and Balance and Symmetry, or Inertia as Newton's first law.

It seems reasonable measuring the properties of the Matrix centre and the current characteristics of an object or a system in the standard units related to the current understanding of space, time, energy, and force.

Different measurement systems must be applied to the currently inaccessible Spaces of the Progressive and Regressive Time and the 2-dimensional Info-Energy grid that forms them.

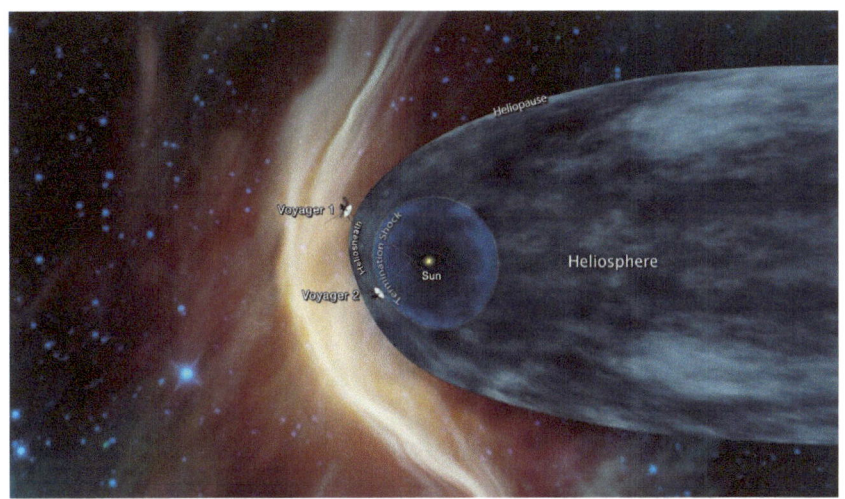

Image 6: 'Voyager Probes Heliosphere Chart' Image credit: NASA/JPL-Caltech

'This graphic shows the position of the Voyager 1 and Voyager 2 probes, relative to the heliosphere, a protective bubble created by the Sun that extends well past the orbit of Pluto. Voyager 1 crossed the heliopause, or the edge of the heliosphere, in 2012. Voyager 2 is still in the heliosheath, or the outermost part of the heliosphere.' NASA

Matrix Balance

Isaac Newton defined inertia as his first law. He described the innate force of matter as 'a power of resisting by which every body, as much as in it lies, endeavours to preserve its present state...' [Newton Isaac, Mathematical Principles of Natural Philosophy (1687)]

Albert Einstein indicated, 'It was found that inertia is not a fundamental property of matter, nor, indeed, an irreducible magnitude, but a property of energy.' [Einstein Albert, A Brief Outline of the Development of the Theory of Relativity (1921)]

The modern understanding interprets inertia as 'a property of matter by which it continues in its existing state of rest or uniform motion in a straight line, unless that state is changed by an external force' (Oxford dictionary).

To sum, inertia as a tendency to remain unchanged or resist to changes is, indeed, a property of energy and matter. Every Matrix displays a tendency to resist any changes in its energy, associated information, and Space-Time. Every existing object and every system of the objects display a tendency to obtain and retain its Space-Time, Info-Energy, and Mass-Energy balance and symmetry through its Space-Time axis and 'o' Space-Time point at the centre of the object or the system's multimodal Space-Time and Info-Energy structure.

Accordingly, the Matrix of an object or a system is symmetrical through the Space-Time axis and 'o' Space-Time point at the Matrix centre if the Matrix is balanced. A mirror effect is possible.

The 'o' time-point, reflecting the human perception of time and representing the time component of the 'o' Space-Time point, indicates the Matrix centre.

The 1-dimensional Space-Time and associated Info-Energy are balanced and symmetrical through the 'o' time-point at the centre of the Matrix if the Matrix is balanced.

The fundamental 2-dimensional Space-Time and associated Info-Energy of the 2-dimensional grid are balanced and symmetrical through the Space-Time axis and the 'o' time-point at the Matrix centre if the Matrix is balanced. Accordingly, the Spaces of the Progressive and Regressive Time are symmetrical through the 'o' time-point if the Matrix is balanced. The Arrows of Time are directed along the Matrix axis, contra-directed, and symmetrical through the 'o' time-point - the point 'now', if the Matrix is balanced that means the related multimodal object or the system is balanced in its Space-Time, Mass-Energy, and Info-Energy.

The human perception of time as a point 'now' provides us with the unlimited actuality of space and space-related objects and systems and makes available our interpretation of the object/system's volume. The object's Space-Time imbalance - relative to the 'o' Space-Time point at the centre of its Matrix, and associated Info-Energy and Mass-Energy imbalance provide us with the Space of the Current Time - the object's volume filled with mass, kinetic energy, and other energy representations acting in the volume of the object in the period of time now occurring.

The potential and actual Info-Energy structures counteract and keep a balance in the Matrix, following the laws of Space-Time, Info-Energy, and Mass-Energy Conservation, Transformations, Reversibility, Limitation, and Balance and Symmetry. The Matrix resultant vector-force equals zero if the Matrix is balanced and symmetrical through the Space-Time axis and 'o' Space-Time point.

In the balanced TRM of a radiating object (Figure 1), the Matrix resultant vector-force reflects the balanced influence of the Matrix grid vector-force of pressure developing the Matrix Spaces of the Progressive and Regressive Time. The balanced TRM of a radiating object displays the balanced tendency to transform the object's volume into the Matrix Spaces of the Progressive and Regressive Time, along with the tendency to transform the object's actual Info-Energy into the potential Info-Energy of the Matrix 2-dimensional grid forming the Spaces of the Progressive and Regressive Time. The TRM is balanced if its Space-Time and Info-Energy are balanced through the centre of the Matrix (Figure 1).

The high energy radiating objects and systems, including such massive bodies and systems as planets, stars, and radiating star clusters and galaxies tangled by gravity, demonstrate the Space-Time and Info-Energy imbalance and balancing transformations.

In the balanced SRM of a non-radiating object (Figure 2), the Matrix resultant vector-force reflects the balanced influence of the object's vector-force of resistance developing the volume of the object in the Matrix Space of the Current Time. The SRM, existing in a dynamic balance, is the typical Matrix of a non-radiating, non-developing vacuum that does not significantly increase in size, such as outer space that is of low density and pressure in our region of the Universe. Its SRM exists in a dynamic balance. Another example is from the quantum world - in quantum field theory, a false vacuum is a hypothetical vacuum that is not entirely stable. The balanced SRM displays the balanced tendency to transform the Matrix Spaces of the Progressive and Regressive Time into the object's volume, along with a tendency to transform the time-associated potential Info-Energy of the Matrix grid into the object's actual Info-Energy. The SRM is balanced if its Space-Time and Info-Energy are balanced through the Matrix centre (Figure 2).

The large spacious non-radiating and Space-Rising objects and systems, such as the non-radiating Space developing star clusters and galaxies, systems building Black Holes, and our Universe, demonstrate a Space-Time imbalance and Space-Time and Info-Energy transformations.

Equilibrium

Internal equilibrium of the multimodal Space-Time and Info-Energy structure of the object or the system is reached if the Matrix is balanced and symmetrical in Space-Time, Info-Energy, and Mass-Energy that means that no space, time, mass, energy, and information enter or leave it, it is balanced under its intrinsic characteristics, symmetrical through the 'o' Space-Time point and the Space-Time axis, and its energy is spatially and temporally uniform. Internal equilibrium of the Matrix occurs when the Matrix resultant force equals zero.

In case of the contact equilibrium between Matrixes, Space-Time, Info-Energy, and Mass-Energy are transferred through the contact paths, and the relation of the equilibrium is transitive, reflexive, and symmetrical. The contact equilibrium is reached if the Matrix of the new-formed system is balanced and symmetrical through the 'o' Space-Time point and the Space-Time axis. It means that

Space-Time, Info-Energy, and Mass-Energy of the system's Matrix are balanced and symmetrical through the 'o' Space-Time point and Space-Time axis, and the resultant force of the system's Matrix equals zero.

A balanced, isolated object or an isolated system in a state of equilibrium would cease to exist. If we apply an understanding of the electric charge to the isolated object or the isolated system in the state of equilibrium, we can see that the properties, forces, and electric charges cancel out, yielding net properties and charge of zero, thus making the object neutral. This description provides an essential insight into a balanced, isolated object or an isolated system in a state of equilibrium.

Planck length, Planck time, and Planck energy are to be understood as the minimal Space-Time and Info-Energy imbalance and, accordingly, the minimal conditions of an object or a system's existence in our dynamic world.

Balancing forces, acting between dissimilar dimensional layers of the Matrix structure, influence the directions of the Space-Time and Info-Energy 'flow' and the reverse of these directions in the existing multimodal objects and systems.

Time Dimensions

We mentioned above that every existing single object and every system of the objects, including visually 'empty' spaces, subatomic particles and holes, systems holding Black Holes, other objects and systems, and the Universe, are multimodal objects and systems carrying the properties of different Space-Time modalities, which are reflected in the structure of their Matrixes involving a coexistence of different modalities and dimensions of Space-Time.

In our daily life, we apply the point-of-time settings, associated with our perception of time as a point 'now', and the linear time settings according to our understanding of time as a timeline with the time 'flow' from the past to the future.

Events of the past and future, located relatively close to the point 'now' in time, are usually more actual and easily designed or remembered, than events distant in time. Nevertheless, we cannot leave psychology aside. Emotional events of the past are full of energy and easily remembered. They are felt actual, 'real' to us and more influential to the current events, or say, their location is relatively closer to the time-point 'now' than events distant in time and long forgotten.

In the first time dimension (Figures 1, 2), an object is located along the Space-Time axis at the Matrix centre associated with the point 'now' in time. It is a typical position of an object (or a system) in its Matrix. Info-Energy blocks, located relatively close to the '0' time-point, are situated closer to the centre of the Matrix, than other latent information.

The application of the first time dimension to the TRM was found experimentally. Works, associated with the Light Cone, describing in great details the temporal evolution of a flash of light in Minkowski Space-Time, were a perfect starting point to work from.

If aiming to analyse a considered Space-Time region, we apply the point-of-time and linear-time dimensions to the object or the system in its multimodal Matrix.

There is no sign of a gravitational field that would generate accelerated motion relative to an observer located outside of the Matrix or an observer situated at the Matrix centre.

Applying the first time dimension to the objects and systems in their Matrixes, we detect the rotation of the volumes of the unbalanced multimodal objects and systems about their Space-Time axis under the influence of the Matrix forces.

The rotation of the object or the system's volume about its Space-Time axis in the multimodal Space-Time is reflected in the specific rotation of this object or the system about its axis of rotation in our dynamic world.

In the second time dimension (Figures 3, 4), the centre of an object or a system is directed to the Matrix Space of the Regressive Time while its surface is situated facing the Matrix Space of the Progressive Time.

The second time dimension reflects the function of the human consciousness on the past edge of the sensory input while developing a mental conception on the base of the sensory data and previously learned information.

Applying the second time dimension to objects and systems in their Matrixes, we detect the Space-Time and Info-Energy imbalance in high energy massive radiating objects and systems and large spacious non-radiating objects and systems. For example, the systems, holding Black Holes, demonstrate the Space-Time and Info-Energy imbalance in the second time dimension facing our dynamic world.

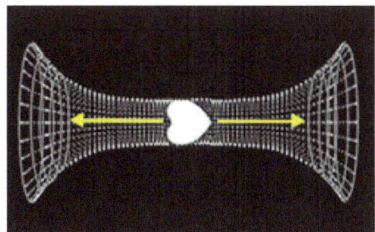

Figure 3: TRM in the second time dimension

The Arrows of Time are not harmonically balanced in the Matrixes of these objects and systems in the second time dimension and 2-dimensional time settings.

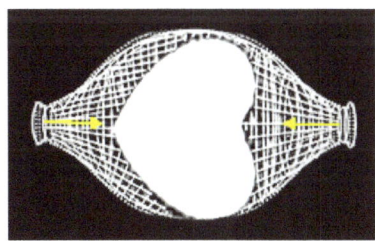

Figure 4: SRM in the second time dimension

The Matrixes display the curved line of the structured representation of the unbalanced function, reflected in the Space-Time and Info-Energy imbalance, gravity, antigravity, and geometry of unbalanced objects and systems.

Besides, applying the second time dimension to an unbalanced system in its Matrix, we detect a new centre of the Matrix symmetry - a new '0' time-point at the centre of the Matrix. Different representations of the same '0' time-point correspond to the realisation of the same Matrix seen in two different time dimensions, such as the TRM in Figures 1 and 3 or the SRM in Figures 2 and 4. We can detect two different representations of the same '0' time-point, which coexist in the 2-dimensional time settings in the multimodal Space-Time of the existing objects and systems.

The coexistence of two different representations of the '0' time-point confirms a coexistence of the qualities of a toroid along with the qualities of a globe in the high energy massive radiating objects, such as our planet, in the Matrix multimodal Space-Time, and accordingly, the development of the Earth geometry as the degeneration of a toroid in our dynamic world. The coexistence of the qualities of a toroid in the past and the qualities of a globe in the future is the cause of the specific geometry of our planet existing in our dynamic world.

However, two different representations of the '0' time-point, existing in the Universe 2-dimensional Time settings, confirm a coexistence of the qualities of a toroid along with the qualities of a globe in the multimodal Space-Time reflecting the evolution of the Universe geometry as the developing toroid in our dynamic world in the future. The coexistence of the qualities of a globe, related to the past of the Universe, and the qualities of a toroid, related to its future, is the cause of the specific geometry of the thermodynamic Universe.

Image 7: 'Omega Centauri' Image credit: NASA/JPL-Caltech/UCLA

'NASA's Wide-field Infrared Survey Explorer, or WISE, has captured a favourite observing target of amateur astronomers - Omega Centauri. Also known as NGC 5139, this celestial cluster of stars can be found in the constellation Centaurus and can be seen by the naked eye to observers at low northern latitudes and in the southern hemisphere. Omega Centauri contains approximately 10 million stars and is about 16,000 light-years away.

'The ancient astronomer Ptolemy thought Omega Centauri was a star, and Edmond Halley identified it as a nebula in 1677. In the 1830s, John Herschel identified it as a globular star cluster orbiting our Milky Way galaxy. A globular cluster is a spherical group of stars that are bound together by gravity.

'Recent research based on observations using NASA's Hubble Space Telescope and the Gemini Observatory indicates that there is a black hole at its center. This suggests that Omega Centauri may actually be a dwarf galaxy that has been stripped of its outer stars and not a globular cluster after all.' NASA

Matrix Imbalance

There is no 'empty' space in the Universe, and no isolated objects and systems exist in our dynamic world. The 'outer space' is always another object or a system. Objects and systems, existing in the Universe, are tangled together by the Space-Time, Info-Energy, and Mass-Energy imbalance. The unbalanced Time-Rising systems, such as high energy massive radiating systems, undergo their natural development by increasing in time. The unbalanced Space-Rising non-radiating systems, such as systems holding Black Holes and our Universe, undergo their natural development by increasing in size.

The Space-Time imbalance and the associated Info-Energy and Mass-Energy imbalance are detectable in the second time dimension (Figures 3, 4) and 2-dimensional Time settings. Applying the second time dimension to the unbalanced, multimodal structure of a system, we detect a new centre of the Matrix symmetry. Two different representations of the '0' time-point, existing in 2-dimensional time settings within the Matrixes of the objects and systems, indicate the Space-Time and Info-Energy imbalance. Besides, the unbalanced Matrixes display the curved line of the structured representation of the unbalanced function that is specific to the related object or the system. The Space-Time, Info-Energy, Arrows of Time, and Arrows of Space are not harmonically balanced in the Matrixes of these systems.

The excessive time, detectable in the second time dimension, is directed from the surface of high energy massive radiating systems (Figure 5). This Space-Time imbalance is supported by the Mass-Energy and Info-Energy imbalance, such as the excessive time-associated latent and potential Info-Energy of the 2-dimensional grid and the deficit of the actual Info-Energy represented in masses, kinetic energy, and other energy representations acting in the peripheral regions of the system in the period of time now occurring. In modern physics, the deficit of mass and kinetic energy of the system is called 'the negative gravitational potential energy'. This Space-Time imbalance and the associated Mass-Energy and Info-Energy imbalance humans perceive as gravity on the surface of high energy massive radiating systems. For example, we can perceive gravity on the surface of the Earth and detect signs of gravity in other high energy massive radiating systems, such as stars and planets.

The deficit of time is directed from the centres of high energy massive radiating systems (Figure 6). This Space-Time imbalance is supported by the Mass-Energy and Info-Energy imbalance, such as the deficit of time-associated latent and potential Info-Energy of the 2-dimensional grid and the excessive actual Info-Energy represented in masses, kinetic energy, and other energy representations acting in the central region of the system in the period of time now occurring. In modern physics, the excessive mass and kinetic energy of the system is called 'the positive mass-energy'. This Space-Time imbalance and the associated Mass-Energy and Info-Energy imbalance are reflected in antigravity. Antigravity, prompted by the centre of the Earth, balances gravity prompted by the peripheral region of the Earth and therefore protects the planet from the gravitational shock. The projection of the different representations of the 'o' time-point, existing in the 2-dimensional time settings, into the volume of the high energy massive radiating system, such as our planet and the Sun, would draw the particular region of the Space-Time and Info-Energy imbalance with the qualities of intense antigravity at the centre of the system.

The Space-Time imbalance and the associated Mass-Energy and Info-Energy imbalance are reflected in the specific geometry of the high energy massive radiating objects and systems in our dynamic world. The two different representations of the '0' time-point, existing in 2-dimensional time settings, confirm the coexistence of the qualities of a toroid along with the qualities of a globe in the multimodal Space-Time and Info-Energy structures of the high energy massive radiating systems. The systems' geometry may be represented as the degeneration of a toroid in our dynamic world. This historical development of our planet is the cause of its Space-Time and Mass-Energy imbalance, reflected in gravity on the surface of the Earth.

In the General Theory of Relativity, Albert Einstein described gravity as a consequence of the curvature of Space-Time caused by the uneven distribution of mass. In astrophysics, an event horizon is a boundary beyond which events cannot affect an observer. In our dynamic world, we operate in the Space of the Current Time. As we sum these statements up, we can introduce the boundary of the curvature of the system's Space-Time as an event horizon that surrounds the total volume of the system that equals the system's Space of the Current Time. As such, the external boundary of the curvature of the Earth Space-Time - affected by the Earth gravity, is the external boundary of the Earth Space of the Current Time. In case of a high energy massive radiating system, such as the Earth, the system's Arrows of Space start at the external boundary of the curvature of the system's Space-Time affected by gravity. In our descriptions, we refer the boundary of the curvature of the system's Space of the Current Time as 'the surface of the system'.

The specific toroid-globe geometry and the associated uneven distribution of mass are the cause of the particular geometry of a high energy massive radiating system, such as the Earth, in our dynamic world.

The Space-Time imbalance, supported by the Mass-Energy and Info-Energy imbalance, specific geometry of the high energy massive radiating systems, and the balancing forces, acting between dissimilar dimensional layers of the system, are reflected in the specific rotation of these systems about their axis of rotation in our dynamic world and the associated rotation of the dynamic representations of background radiation that must be detectable. We will talk about the influence of the balancing forces on the rotation of the system in the next chapter.

However, Space-Rising non-radiating systems also demonstrate the Space-Time, Mass-Energy, and Info-Energy imbalance detectable in the second time dimension and 2-dimensional time settings.

The deficit of time is directed to the surface of large spacious non-radiating systems (Figure 7), such as toroidal systems, systems holding Black Holes, and our dynamic Universe. This Space-Time imbalance is supported by the Mass-Energy and Info-Energy imbalance, such as the deficit of time-associated latent and potential Info-Energy of the 2-dimensional grid and the excessive actual Info-Energy represented in masses, kinetic energy, and other energy representations acting in the peripheral region of the system in the period of time now occurring. We mentioned above that in modern physics, the excessive mass and kinetic energy of the system is called 'the positive mass-energy'. This Space-Time imbalance and the associated Mass-Energy and Info-Energy imbalance are reflected in antigravity in the peripheral regions of the Space-Rising non-radiating systems.

Processes, reflected in antigravity in the Universe, impact on human concepts of 'limitless' or 'endless' in space, extent, and size; mathematical concepts of transfinite and infinite numbers, and ideas corresponding to the infinite Universe and the infinite number of stars in the Universe that is 'impossible to measure or calculate'. It reminds me of the time before Leif Erikson discovered 'Vinland'.

The excessive time is directed to the centres of the spacious non-radiating systems (Figure 8). This Space-Time imbalance is supported by the Mass-Energy and Info-Energy imbalance, such as the excessive time-associated latent and potential Info-Energy of the 2-dimensional grid and the deficit of the actual Info-Energy represented in masses, kinetic energy, and other energy representations acting in the central region of the non-radiating system (sometimes referred as 'the zero-point radiation of the vacuum' of Feynman and Wheeler) in the period of time now occurring. We mentioned above that in modern physics, the deficit of mass and kinetic energy of the system is called 'the negative gravitational potential energy'. This Space-Time imbalance and the associated Mass-Energy and Info-Energy imbalance are reflected in gravity at the centres of the spacious, Space-Rising and toroidal non-radiating systems frequently holding the central transmitting Black Holes. A Black Hole is a contact path equilibrating the system. The centres of large

Space-Rising non-radiating systems display the qualities of intense gravity in the second time dimension facing our dynamic world. Nasa scientists observe and photograph Black Holes by detecting their effect on the matter nearby.

The projection of the different representations of the '0' time-point, existing in the 2-dimensional time settings, into the volume of an unbalanced Space-Rising non-radiating system, such as a system holding the central Black Hole or the self-rising vacuum, would draw the particular region of the Space-Time and Info-Energy imbalance with the qualities of intense gravity at the centre of the system. The volumes, energy, and matter, absorbed through the transmitting Black Holes, are being processed and transformed into the forms of space, energy and matter of the particular spacious Space-Time region, which this Black Hole develops. A Black Hole evaporates if the multimodal non-radiating system is balanced and the resultant force equals zero.

The Space-Time imbalance and the associated Mass-Energy and Info-Energy imbalance are reflected in the specific geometry of the large spacious and Space-Rising non-radiating systems in our dynamic world. The two different representations of the '0' time-point, existing in the 2-dimensional time settings, confirm a coexistence of the qualities of a toroid with the qualities of a globe in Space-Rising non-radiating systems, such as the thermodynamic Universe, in the multimodal Space-Time. Their specific toroid-globe geometry reflects the developing toroidal form of these systems in our dynamic world in the future.

The Space-Time imbalance, supported by the Mass-Energy and Info-Energy imbalance, specific geometry of the Space-Rising non-radiating systems, and the balancing forces, acting between dissimilar dimensional layers of the system, are reflected in the specific rotation of these systems about their axis of rotation in our dynamic world and the associated rotation of background radiation that must be detectable. We will talk about the influence of the balancing forces on the rotation of the system in the next chapter.

The rotation of the centre of the non-radiating system is currently detectable - it influences the rotation of the matter before it falls onto a Black Hole. Other aspects of the non-radiating and toroidal objects and systems, developing the 2-dimensional potential space and accumulating tremendous potential energy resources (as Feynman and Wheeler mentioned), are currently unknown to us. Their latency is associated with the limits of our perception of the potential space, time, and the associated potential Info-Energy.

The Equivalence principle

The Equivalence principle of General Relativity states that 'at any point of Space-Time the effects of a gravitational field cannot be experimentally distinguished from those due to an accelerated frame of reference' (Oxford dictionary).

We, supporting the Equivalence principle of General Relativity, view the mentioned gravitational field as the Space-Time imbalance, such as the excessive time and deficit of space supported by the associated Info-Energy and Mass-Energy imbalance, which humans perceive as gravity. At any point of Space-Time, the effects of the specific Space-Time imbalance, such as the excessive time and deficit of space, supported by the Info-Energy and Mass-Energy imbalance and reflected in gravity, cannot be experimentally distinguished from those due to an accelerated frame of reference.

We, developing the Equivalence principle of General Relativity, view antigravity as the specific Space-Time imbalance, such as the excessive space and deficit of time supported by the Info-Energy and Mass-Energy imbalance. At any point of Space-Time, the effects of the specific Space-Time imbalance, such as the excessive space and deficit of time, supported by the Info-Energy and Mass-Energy imbalance and reflected in antigravity, cannot be experimentally distinguished from those due to a decelerated frame of reference.

Matrix Forces

The objects and systems, existing in our dynamic world, are tangled together by the Space-Time, Info-Energy, and Mass-Energy imbalance.

The excessive time and space deficit, supported by the Info-Energy and Mass-Energy imbalance, are detectable on the periphery of the high energy massive radiating objects and systems and at the centres of the spacious non-radiating objects and systems. We can perceive the Earth gravity. We can detect gravity of Black Holes when we observe their effects on the nearby matter. We can observe gravity when the particles and energy appear in a vacuum.

The excessive space and time deficit, supported by the Info-Energy and Mass-Energy imbalance, are detectable at the centres of the high energy massive radiating objects and systems and on the periphery of the non-radiating objects and systems. We cannot perceive the Earth antigravity directly. We can detect antigravity as the cosmological redshift and the expansion of the Universe. We observe antigravity when the particles and energy disappear in a vacuum.

Every Matrix demonstrates a tendency to resist any changes. The unbalanced multimodal Space-Time and Info-Energy structures display a tendency to obtain and retain a balance and symmetry through its Space-Time axis and 'o' Space-Time point at the centre of the Matrix. The Matrix imbalance resolves itself in the balancing transformations. Similarly, gravity, exposing Space-Time, Info-Energy, and

Mass-Energy imbalance, results in gravitational acceleration, and antigravity is settled in anti-gravitational deceleration.

Balancing transformations are initiated by the influence of the two opposite vector-forces acting in a dynamic balance at the Matrix centre - the Matrix grid vector-force of pressure and the system's vector-force of resistance.

Balancing forces emerge within the fundamental 2-dimensional Space-Time. The Matrix grid vector-force of pressure diverges from the object or the system's vector-force of resistance within the 2-dimensional grid of the potential Info-Energy arranged by background radiation. Info-Energy blocks, processed, stored, and transmitted by the 2-dimensional grid, alter the object or the system's structures and functioning in a manner specific to the object or the system.

The Matrix grid vector-force of pressure applies pressure (contact force) on the surface of the system embedded at the centre of the Matrix. This vector-force has a direction perpendicular to the surface of contact (normal force). This pressure is a measure of the system's latent, or potential, Info-Energy stored in the Matrix grid. It is related to the latent energy density and latent information density associated with the system's potential space and time of existence. The Matrix grid vector-force of pressure limits the system's volume in the Matrix Space of the Current Time and develops the system's Spaces of the Progressive and Regressive Time.

The influence of the grid vector-force of pressure upon the system's volume is associated with the Matrix tendency to transform the Space of the Current Time, or the volume of the system, into the Matrix Spaces of the Progressive and Regressive Time, along with the tendency to transform the actual Info-Energy, acting in the volume of the system as mass and kinetic energy, into the potential Info-Energy of the 2-dimensional grid, retaining the total Space-Time, Mass-Energy, and Info-Energy of the Matrix unchanged following the laws of Space-Time, Info-Energy, and Mass-Energy Conservation, Transformations, Reversibility, Limitation, and Balance and Symmetry.

If an object or a system has reached the limits, such as Planck length, Planck time, and Planck energy, actuating the reverse of the Space-Time and Info-Energy 'flow', it is being reversed by the Matrix forces.

The system's actual Info-Energy, bounded with the Matrix grid in the Space of the Current Time, results in the system's vector-force of resistance, which influences the 2-dimensional grid (contact force). This vector-force has a direction perpendicular to the surface of contact (normal force). The system's vector-force of resistance develops the Space of the Current Time. The vector-force of resistance is a measure of the system's actual Info-Energy acting in the volume of the system in the period of time now occurring.

The influence of the system's vector-force of resistance upon the Matrix 2-dimensional grid is associated with the Matrix tendency to transform the Spaces of the Progressive and Regressive Time into the Space of the Current Time, or the volume of the system, along with the tendency to transform the latent, or potential, Info-Energy of the 2-dimensional grid into the system's actual Info-Energy, retaining the total Space-Time, Mass-Energy, and Info-Energy of the Matrix unchanged under the laws of Space-Time, Info-Energy, and Mass-Energy Conservation, Transformations, Reversibility, Limitation, and Balance and Symmetry.

The influence of the Matrix forces leads to the specific rotation of these objects and systems about their Space-Time axis. They initiate the rotations in opposite directions with an unequal resultant quantity and impulse of the rotation of a system. Accordingly, the resultant quantity of rotation of a system and the impulse of the rotation are unequal zero. The specific rotation of the system about its Space-Time axis in the multimodal Space-Time is reflected in the associated rotation of the system about its axis in our dynamic world.

The relative specific rotation of the Matrix grid is reflected in the rotation of background radiation, such as cosmic background radiation. The relative rotation of cosmic background radiation, being at the same time the external framework of the Universe in the Matrix and the internal dynamic skeleton of the Universe and the existing objects and systems in our dynamic world, must be detectable.

The Matrix grid vector-force of pressure and the object's vector-force of resistance interact in a dynamic balance. Dynamics preserve the laws of Space-Time, Info-Energy, and Mass-Energy Conservation, Transformations, Reversibility, Limitation, and Balance and Symmetry, or Inertia as Newton's first law.

The influence of the Matrix resultant vector-force reflects the dominant influence of the Matrix grid vector-force of pressure or the object's vector-force of resistance. The influence of the Matrix resultant vector-force provides the balancing Space-Time, Info-Energy and Mass-Energy transformations. The Matrix resultant contact force equals zero if the Matrix is balanced.

Time and Energy Flow

The excessive time and time-associated potential Info-Energy and deficit of actual space and masses are easily detectable as gravity in the peripheral regions of high energy massive radiating objects and systems, such as Globular star clusters and galaxies tangled by gravity, stars and planets.

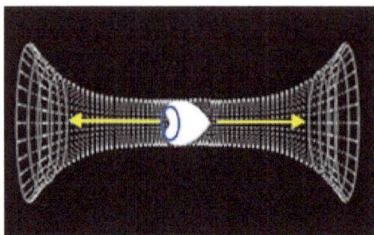

Figure 5: TRM Gravity

The Arrows of Time are not harmonically balanced within these systems. They indicate the current direction of time and time-associated Info-Energy 'flow' within a high energy massive radiating system. The excessive time and time-associated potential Info-Energy are directed from the periphery of the high energy massive radiating objects and systems (Figure 5).

The deficit of the volume, masses, and kinetic energy in the peripheral regions of a high energy massive radiating system (Figure 5) impact the direction of time and time-associated latent Info-Energy 'flow' within the system. The system's vector-force of resistance, prompted by the periphery of the high energy massive radiating system, transforms the system's excessive time and time-associated potential Info-Energy into the actual space, kinetic energy, and masses of the high energy massive radiating system, such as our planet. The balancing Space-Time, Mass-Energy, and Info-Energy transformations are reflected in gravitational acceleration prompted by the periphery of the system.

High energy massive radiating objects and systems keep the special secret properties at their centres (Figure 6).

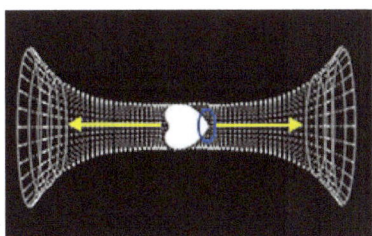

Figure 6: TRM Antigravity

The excessive time and time-associated latent Info-Energy on the periphery of the high energy massive radiating system coexist in a dynamic balance with the excessive space and the associated actual Info-Energy at the system's centre. The Arrow of the Regressive Time, directed along the Space-Time axis, indicates the direction of time and time-associated Info-Energy 'flow' from the centre of the high energy massive radiating system (Figure 6).

The excessive space, mass, and kinetic energy at the centres of the high energy massive radiating systems coexist with the deficit of time and time-associated potential Info-Energy reflected in antigravity. The excessive space, masses, and kinetic energy at the centre of a high energy massive radiating system influence the direction of the time and time-associated energy 'flow'. The Matrix grid vector-force of pressure, prompted by the central regions of a high energy massive radiating system, transforms the system's excessive space, mass, and kinetic energy into the system's time, potential space, and associated potential Info-Energy. The balancing Space-Time and Info-Energy transformations are reflected in anti-gravitational deceleration prompted by the centre of the high energy massive radiating system (Figure 6).

The projection of the two different representations of the '0' time-point, coexisting in the 2-dimensional time settings within the system's multimodal structure, into the volume of a high energy massive radiating system would draw the particular region of intense antigravity at the centre of the system and possible location of the transmitting Black Hole at the centre of a high energy massive radiating system, such as a star or a radiating galaxy tangled by gravity.

Antigravity, prompted by the centres of the high energy massive radiating systems, balances gravity prompted by their peripheral regions and therefore protects these systems from the gravitational collapse. Antigravity existence is currently rejected as 'impossible'. The fact of antigravity existence must not be dismissed. Antigravity must be investigated. To make it simple, we can think about the observer who had reached the centre of the Earth. Could he continue moving deeper and more profound? No, he couldn't. He would be stopped by the 'natural limit' - the impossibility of moving further to the centre. It would be similar to reaching the pointed top of a mountain.

Antigravity is the most intense on the periphery of the Universe (Figure 7). Anti-gravitational processes on the periphery of the Universe do not contradict but prompt the local anti-gravitational processes in different areas of the Universe.

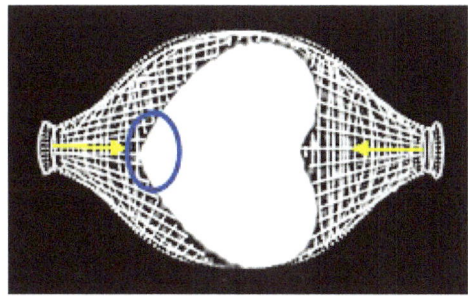
Figure 7: SRM Antigravity

The deficit of time and time-associated potential Info-Energy are directed to the peripheral areas of the large unbalanced non-radiating systems, such as the Space-Rising galaxies, toroidal systems, and the Universe. Arrows of Time are not harmonically balanced within the Space-Rising and toroidal systems (Figure 7). Arrows of Time, directed along the Space-Time axis, indicate the direction of time and time-associated Info-Energy 'flow' within the system. The excessive space and space-associated actual Info-Energy affect the direction of time and time-associated energy 'flow' in the multimodal structure of the non-radiating system.

The influence of the Matrix grid vector-force of pressure, prompted by the peripheral regions of these systems, transforms the excessive actual space, masses, and kinetic energy into the potential space, time, and the associated potential Info-Energy of the system. The balancing Space-Time and Info-Energy transformations are reflected in anti-gravitational deceleration prompted by the periphery of the non-radiating system (Figure 7).

The deficit of time and time-associated potential Info-Energy on the periphery of the large Space-Rising non-radiating systems coexists in a dynamic balance with the excessive time and time-associated latent Info-Energy at the centres of the systems (Figure 8).

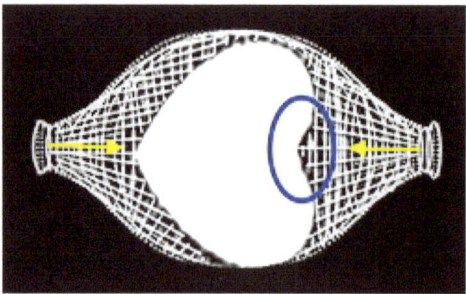

Figure 8: SRM Gravity

The excessive time and time-associated Info-Energy are directed to the centre of an unbalanced spacious and Space-Rising non-radiating system. The deficit of the volume at the centre of the system accompanies the deficit of masses, kinetic energy and other energy representations acting at the centre of the system in the period of time now occurring. This Space-Time imbalance and the associated Mass-Energy and Info-Energy imbalance, which are reflected in gravity, influence the direction of the time and time-associated energy flow within the multimodal non-radiating system. The Arrow of the Regressive Time is directed to the centre of a non-radiating system. It indicates the time and time-associated energy flow prompted by the centre of the system (Figure 8).

The influence of the system's vector-force of resistance, prompted by the centre of the system, transforms the excessive time and time-associated latent Info-Energy into the volume, mass, and kinetic energy of the system. The balancing Space-Time and Info-Energy transformations are reflected in gravitational acceleration prompted by the centre of the non-radiating system. The deficit of the volume, energy, and mass at the centre of the non-radiating system might initiate the formation of the transmitting Black Hole delivering the volumes, masses, and energy to the centre of this non-radiating system.

The projection of the different representations of the 'o' time-point into the volume of the Space-Rising non-radiating system would draw the particular region of the Space-Time and Info-Energy imbalance with the qualities of intense gravity at the centre of the system and possible location of the system's central Black Hole.

Gravity is the most intense at the centre of the Universe. Gravitational processes at the centre of the Universe do not contradict but prompt the local gravitational processes in different areas of the Universe.

The balancing forces, acting between dissimilar dimensional layers of the system's multimodal structure, balance the system's existing capacity to perform work against its actual realisation in the forms of actual energy represented in mass, kinetic energy, and other energy representations acting in the system's volume in the period of time now occurring.

The balancing forces, acting between the multimodal systems within the same layer, build the lines of the force integrating objects into the system and systems into the larger systems. They form the binary and multi-systems to obtain a balance.

The objects and systems, involved in a binary or a multi-system construction, might be distant in space. They are connected within their 2-dimensional Space-Time areas and central points of symmetry of each Matrix. For example, a non-radiating system, receiving volumes, masses, and energy from a massive radiating system through the central Black Hole, is a part of the binary system (Figure 9).

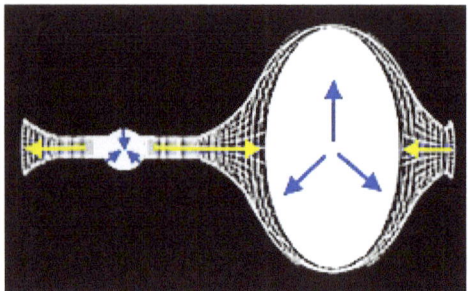

Figure 9: A Space-Rising Black Hole, TRM and SRM

The transmitting Black Hole 'gravitates' and absorbs volumes, matter, and energy from the outer space if the Space-Time imbalance and the associated Info-Energy and Mass-Energy imbalance at the centre of the maternal non-radiating system reflect the excessive time and time-associated latent Info-Energy, along with the deficit of the actual space, mass, kinetic energy, and other energy representations acting in the volume of this system in the period of time now occurring. The maternal non-radiating system, such as a toroidal system, holding the central Black Hole, can receive the volumes, matter, and energy directly from the central regions of the high energy massive radiating system, such as a white dwarf, through the transmitting Black Hole - the direct Space-Time and Info-Energy transfer between Matrixes of these associated systems within the binary system (Figure 9)

Alternatively, the non-radiating system might receive Space-Time and Info-Energy transfer indirectly through the binary system's shared area of the 2-dimensional grid that is arranged by the 2-dimensional representations of background radiation such as cosmic background radiation. The 2-dimensional grid acts as a membrane transforming the absorbed volumes, matter, and energy into the potential space and space-associated potential energy of the shared 2-dimensional Space-Time.

The development of the binary system is determined by the structure of the Space-Time and Info-Energy imbalance between the associated multimodal systems.

Is there a practical application of the Theory of Matrix? Please consider an example.

Creating a binary system with two centres and using the direct Space-Time and Info-Energy transfer between the connected TRM and SRM, we could manage the energy of an atom and create the perfect fuel. The emission of ionising radiation and jets of radiating energy may be released as a result of the spontaneous disintegration of a particle or an atomic nucleus and its 2-dimensional Info-Energy grid at the '0' time-point centre as a consequence of the tremendous excessive actual Info-Energy of the TRM.

It is possible to avoid the spontaneous disintegration of the Matrix centre and manage the energy of this multimodal object by building a binary system with two centres and using the direct Space-Time and Info-Energy transfer between the centres of two Matrixes - the strong TRM of the particle, atom or nucleus with excessive actual Info-Energy and the non-radiating microscopic hole associated with the strong SRM, by analogy with the binary systems in the macro-world, which are usually detectable the 'star and Black Hole' system.

Visually 'empty' spaces are obvious but weak non-radiating systems (weak SRMs) and do not serve the purpose. A self-rising vacuum or a self-rising quantum vacuum could satisfy the requirements for the robust unbalanced Space-Rising non-radiating system. The widening, self-rising quantum vacuum would create the balancing structure and provide a micro-hole for this purpose. Despite a likelihood of being filled with the dark matter and dark energy, the widening, self-rising vacuum would contain the central transmitting region in the second time dimension facing our dynamic world. Accordingly, energy can be taken under control and kept within the relatively balanced binary system. How to recognise the self-rising vacuum? The self-rising vacuum might have an evident displacement of spectral lines toward longer wavelengths - redshift. We suppose that the intensity of displacement could be a sign of the system's ability to receive more energy from its binary partner.

Balancing Transformations

The Space-Time, Info-Energy, and Mass-Energy imbalance and the balancing forces, acting between dissimilar dimensional layers of the system's multimodal structure, impact the direction of the time and time-associated energy 'flow' within the multimodal Space-Time and Info-Energy structure of the system.

The balancing Space-Time and Info-Energy flow is the divergence between the unbalanced elements of the system's different modalities, dissimilar dimensional layers within the system's multimodal Space-Time and Info-Energy structure.

Following the balancing transformations, we start their description from the effects related to the centre of the multimodal Space-Time and Info-Energy structure of a system.

The primary effects, related to the quantum space, quantum time, and quantum energy, are associated with the 'o' Space-Time point at the Matrix centre.

Quantum is a discrete quantity of energy proportional in magnitude to the frequency of the radiation it represents, or say, the energy of each quantum is directly proportional to the frequency of the radiation it represents,

$E = h\nu$,

where E is the energy of each quantum, h is the Planck's constant and ν is the frequency of the radiation.

For example, a photon is the quantum of electromagnetic radiation and the basic unit of light. It can be defined as the quantum of energy. A photon as a quantum of light and other forms of background radiation carries energy proportional in magnitude to the frequency of the radiation it represents. A photon has zero, or to be accurate, currently undetectable, rest mass. The photon's Space-Time structure is reflected in the wavelength, or the oscillations, of the radiation it represents in our dynamic world.

A quantum of energy as a discrete packet of energy can be absorbed, or transformed from the form of a particle and actual Info-Energy into the form of the latent Info-Energy of the Matrix grid - an infinitely thin filament representing the 2-dimensional Space-Time and Info-Energy structure of the system.

A quantum of energy can be released, or say, the form of the latent Info-Energy of the Matrix grid can be transformed into the form of an actual Info-Energy acting as a quantum or a particle in our dynamic world.

According to the Theory of Matrix Natural Laws, the '0' point of energy is not achievable for a system existing in our dynamic world. Planck length, Planck time, and Planck energy reflect the minimal conditions of an object or a system's existence in our dynamic world.

The balancing quantity of flux does not emanate from the '0' space-time-energy point but diverges out of the points of symmetry of the dissimilar dimensional layers of the multimodal Space-Time and Info-Energy structure of a system. Similarly, +x requires -x to establish zero.

The points of symmetry of the dissimilar Space-Time dimensional layers, balancing the multimodal Space-Time and Info-Energy structure of a system, are the points of the Space-Time transformations, supported by Info-Energy, and Mass-Energy transformations, keeping time and space components in a dynamic balance.

The 1-dimensional Space-Time, supported by energy, is an undifferentiated existence of space, time, and energy underlying our dynamic world. It is built by and filled with the one quality Info-Energy of infinite duration, presenting our dynamic world with the ultimate dynamics in accordance with the human perception of the relation of 1-dimensional space to 1-dimensional time as speed.

The 1-dimensional Space-Time and Info-Energy layer is undifferentiated time, space, and energy underlying the system's multimodal structure. The Space-Time and Info-Energy flow is not progression neither emission but the divergence of the coexisting time and space components supported by Info-Energy within the 1-dimensional structure of a system.

The detectable balancing forces emerge within the fundamental 2-dimensional Space-Time. We mentioned above that the system's 2-dimensional structure contains the system's latent Info-Energy. We cannot perceive any movement neither modifications of the 2-dimensional Info-Energy grid forming the Spaces of Time. The system's past appears to be kept merely in memories and memoirs, and we cannot perceive or measure any potential event directly at any distance from the point 'now'. The latent Info-Energy of a system represents the system's capacity to perform work. When the capacity to perform work is being realized, energy is associated with space. It takes perceptible forms.

The Matrix 2-dimensional grid is an active element performing the process of Space-Time, Info-Energy, and Mass-Energy transformations. It carries energy and associated information in the form of Info-Energy blocks in a manner specific to the system. The system's potential space, shaped by the 2-dimensional grid, acts as a membrane, storing the system-specific data and enclosing the volume of the system in the Matrix Space of the Current Time.

The system's potential space is a buffer zone between the opposing forces acting in a dynamic balance. The balancing quantity of flux diverges across the 2-dimensional Info-Energy grid within the system's potential space.

The 2-dimensional grid is the transmitter between the system's time-associated latent Info-Energy and space-associated actual Info-Energy. It operates similar to a filter suppressing and altering the system's structures, processing, keeping, transforming, and transporting energy and associated information in compliance with the stored data to balance and enhance an existing capacity to perform work against its actual realization in forms of actual Info-Energy. Changes of the frequency of electromagnetic waves and their wavelength influence the Matrix grid, changing Space-Time, Info-Energy, and Mass-Energy properties of the associated system through the energy transfer. Accordingly, the balancing flow of Space-Time and Info-Energy diverges within the system's 2-dimensional Space-Time and Info-Energy structure.

Some amount of actual Info-Energy might be transformed into the proportional amount of potential Info-Energy, and some amount of potential Info-Energy may be transformed into the proportional amount of actual Info-Energy. Some amount of space might be transformed into the proportional amount of time, and some amount of time may be transformed into the proportional amount of space. Space-Time transformations are the function of energy. Dynamics preserve the Theory of Matrix Natural Laws.

The multimodal Space-Time and Info-Energy structures provide a mechanism for the transmission of balancing transformations in our dynamic world. The 2-dimensional representations of background radiation arrange the system's 2-dimensional grid. It is the 'gravitational horizon' and 'anti-gravitational horizon', or say, 'transformations horizon' in the process of the balancing Space-Time, Mass-Energy, and Info-Energy transformations, including those associated with gravity and antigravity.

The interrelationship between the 2-dimensional representations of background radiation and their dynamic representations, arranging the internal dynamic skeleton of the objects and systems in our dynamic world, provides a mechanism for the transmission of the Space-Time, Info-Energy, and Mass-Energy transformations in our dynamic world. The frequency of the electromagnetic waves and their wavelength influence the Matrix grid, changing Space-Time properties through the energy transfer.

Accordingly, the flow of Space-Time and Info-Energy within the object's multimodal structure affects the Space-Time and Info-Energy flow within the volume of the object in our dynamic world. The effects of the balancing Space-Time and Info-Energy flow are detectable as changes of volume, mass, kinetic energy, shape and structure, magnetic fields and electric currents. It influences the intensity of gravity and antigravity, acceleration and deceleration, and accelerates and decelerates other qualitative and quantitative changes within the existing objects and systems.

The potential and actual Info-Energy structures co-interact and keep a balance within the system's multimodal Space-Time. Similarly, the multimodal structures of the joined objects and systems interact and maintain a balance within the larger systems. As a result, we can observe the binary or multi-systems of galaxies and Black Holes - the centres of the non-radiating systems in the second time dimension facing our dynamic world, and their integration. The Hawking radiation is a sign of the Matrix grid destruction at the Matrix centre or in the area of the Matrixes' connection, if the 'o' time-point centre was built, as a consequence of the tremendous excessive Info-Energy between connected Matrixes.

Objects and systems, existing in the Universe, are tangled together by Space-Time and Info-Energy imbalance and balancing transformations. In our dynamic world, the Space-Time, Info-Energy, and Mass-Energy transformations propagate with the speed of light. They are carried by background radiation.

The Theory Of Light

'Relativity theory ... shares with the corpuscular theory of light the unusual property that light carries inertial mass from the emitting to the absorbing object.' [Einstein Albert, The Development of Our Views on the Composition and Essence of Radiation (1909)]

The Theory of Matrix develops the theory of light, uncovers unusual property of light and introduces the speed of light as the coefficients [c] and $[c]^2$ applicable to the decisions associated with Space-Time, Info-Energy, and Mass-Energy transformations and transmissions, including those reflected in gravity and antigravity propagation.

The 2-dimensional grid, arranged by the 2-dimensional representations of background radiation, is being simultaneously the external framework and boundary of a system in the multimodal Space-Time and its dynamic internal skeleton in our dynamic world. It builds the lines of force in the body of the Universe and manages the balancing Space-Time, Mass-Energy, and Info-Energy transformations between the existing objects and systems.

The dynamic representations of background radiation carry Space-Time, Info-Energy, and Mass-Energy transformations with the speed of light in vacuum. The balancing Space-Time, Info-Energy, and Mass-Energy transformations, including those reflected in gravitational acceleration, integrate objects into systems, and the systems into the larger systems.

Background radiation and light carry inertial mass from the emitting object to the absorbing object within the system and from the emitting system to the absorbing system within the larger system, integrating the existing objects and systems into the multimodal Universe.

Simultaneously, the balancing Space-Time, Info-Energy, and Mass-Energy transformations, including those reflected in anti-gravitational deceleration, keep the objects and systems apart as separate entities existing in our dynamic world.

Humans experience qualities associated with the intermediate region of the Universe, and local gravitational and anti-gravitational processes must be detectable.

Take a specific example and consider the Hubble's law and the relationship between galaxies of the Local Group - the cluster that includes the Milky Way and more than 50 galaxies, most of which are tiny dwarf systems. Gravitational interactions of galaxies of the Local Group coexist in a dynamic balance with the antigravity in the peripheral out-stretched areas. The Universe comes into sight as expending in accelerated rates. The experimental data appear to us as cosmological redshift indicating the expansion of the Universe - the light is being stretched with redshift being approximately proportional to the galaxy's distance.

The alternative is clear - the cosmological redshift can appear, because the surface of the non-radiating 2-dimensional bubble, containing the galaxies of the Local Group, separates it from the field of the Universe and exhibits anti-gravitational deceleration in increasingly decelerated rates to become the 2-dimensional toroidal system in the future. Accordingly, the Universe appears expanding in increasingly accelerated rates. The area of interest is the surface between the cosmological blueshift and redshift - this area prompts antigravity.

Space-Time equivalence

Albert Einstein has specified the connection between mass and energy of the object, 'If an amount of energy E be given to a body, the inertial mass of the body increases by an amount E/c^2, where c is the velocity of light in vacuo. On the other hand, a body of mass m is to be regarded as a store of energy of magnitude mc^2.' [Einstein Albert, A Brief Outline of the Development of the Theory of Relativity (1921)]. Reading the famous equation for mass-energy equivalence $E = mc^2$ with respect to the 2-dimensional Space-Time helps consider the Theory of Matrix.

We have specified the correlation between the fundamental 2-dimensional time and the fundamental 2-dimensional space of the existing objects and systems. The universal proportionality exists between corresponding amounts of the 2-dimensional time and 2-dimensional space.

Energy is associated with 2-dimensional time the same way as mass, corrected with the squared numerical value of the speed of electromagnetic waves propagation in a vacuum, is associated with 2-dimensional space.

$Et^2 = [c]^2 ml^2$,

where l is length, l^2 is the 2-dimensional space, t is time, t^2 is the 2-dimensional time, E is energy, m is mass, and $[c]^2$ is the Coefficient of Transformation.

Rest mass and rest energy remain proportional to one another the same way as the 2-dimensional space and 2-dimensional time remain proportional to one another.

The universal proportionality exists between equivalent amounts of time associated with energy on the one hand and space associated with mass corrected with the squared numerical value of the speed of electromagnetic waves propagation in a vacuum on the other.

The volume of a system may change, although the total Space-Time of the system remains constant. The total mass of a system may change, although the total energy of the system remains constant. Space-Time and Info-Energy of the system are conserved in the system's multimodal Space-Time and Info-Energy structure.

The Coefficient of Transformation $[c]^2$

High energy massive radiating objects and systems and Space-Rising non-radiating objects and systems undergo the balancing Space-Time, Mass-Energy, and Info-Energy transformations, initiated by the influence of the Matrix forces. These transformations are reflected in the reorganisation of space, time, matter, and energy.

The universal proportionality factor between corresponding amounts of energy and mass equals the speed of light squared. The universal proportionality factor between equivalent amounts of energy and mass, taken as the squared numerical value of the speed of electromagnetic waves propagation in a vacuum, is the coefficient for Space-Time, Mass-Energy and Info-Energy Transformations.

The Coefficient of Transformation $[c]^2$ equals the squared numerical value of the speed of electromagnetic waves propagation in a vacuum. It is the coefficient for Space-Time, Mass-Energy, and Info-Energy transformations and, accordingly, the gravity and antigravity propagation.

The Coefficient of Transformation $[c]^2$ applies to the decisions associated with the Space-Time, Info-Energy, and Mass-Energy transformations. The application of the

Coefficient of Transformation [c]² requires the following formulations.

Transformations

The Space-Time, Mass-Energy, and Info-Energy transformations, related to the generation of space and mass and reflected in gravity propagation, are as following:

A large amount of time is required in order to get a small amount of space proportional to time if the Mass-Energy balance of the object is unchanged. The Space-Time transformation is the function of energy.

A large amount of potential Info-Energy is required in order to get a small amount of actual Info-Energy proportional to potential Info-Energy if the Space-Time balance of the object is unchanged.

A large amount of energy is required in order to get a small amount of mass proportional to energy if the Space-Time balance of the object is unchanged.

Space-Time, Mass-Energy, and Info-Energy transformations, related to the degeneration and reduction of space and mass and reflected in antigravity propagation, are as following:

A small amount of space is required in order to get a large amount of time proportional to space if the Mass-Energy balance of the object is unchanged. The Space-Time transformation is the function of energy.

A small amount of actual Info-Energy is required in order to get a large amount of potential Info-Energy proportional to actual Info-Energy if the Space-Time balance of the object is unchanged.

A small amount of mass is required in order to get a large amount of energy proportional to mass if the Space-Time balance of the object is unchanged.

The balancing transformations, carried by background radiation and light, propagate with the speed of light in a vacuum. The value of the speed of light in a vacuum (c) is 299 792 458 m s-1 (meter per second). Under the Committee on Data for Science and Technology (CODATA) Fundamental Physical Constants, the speed of light in a vacuum (c) is a CODATA Fundamental Physical Constant associated with the velocity of light measurements carried out by Michelson and his associates over the period 1924 to 1935. [NIST. Fundamental Physical Constants (2013)]

The Space-Time Coefficient [c]

We mentioned above that the dynamic representations of background radiation and light constitute the dynamic body of the Universe. The 2-dimensional grid, arranged by background radiation, forms the internal Info-Energy skeleton of the Universe and the existing objects and systems.

The concept of relativity and Einstein's Special Theory of Relativity state that all motion is relative and that the velocity of light in a vacuum has a constant value which nothing can exceed. We add that the world dynamics reflects the asymmetry in our perception of space and time.

The speed of light in vacuum sets the fundamental relation between 1-dimensional space and 1-dimensional time that defines the exact proportion of space to time as the upper limit of our dynamic world. Planck length, Planck time, and Planck energy, based on the calculations using the speed of light in vacuum as the fundamentals proven by experiments, define the exact lower limits for the existing objects and systems and the Universe as we measure and sense it.

Speed as the rapidity of movement represents the relation between 1-dimensional space and 1-dimensional time. The Space-Time Coefficient is a factor that measures the exact proportion of this relation.

The Space-Time Coefficient [c] equals the numerical value of the velocity of light in a vacuum and, accordingly, it is the numerical value of the speed of electromagnetic waves propagation in a vacuum.

The Space-Time Coefficient [c] applies to the decisions associated with the transmission of the Space-Time, Mass-Energy, and Info-Energy transformations, including those reflected in gravity and antigravity propagation in our dynamic world.

The speed of gravity propagation was introduced by Hendrik Lorentz (1900) and confirmed by Hermann Minkowski (1908) as follows, 'The law of mass attraction, which has been just described and which is formulated in accordance with the relativity postulate, would signify that gravitation is propagated with the velocity of light.' [Minkowski Hermann, The Fundamental Equations for Electromagnetic Processes in Moving Bodies, Appendix (1908)]

In our dynamic world, the Space-Time, Info-Energy, and Mass-Energy transformations, including those reflected in gravitational acceleration and anti-gravitational deceleration, propagate with the speed of light. Background radiation and light carry inertial mass from the emitting system to the absorbing system.

Dynamic representations of background radiation, including those currently known as cosmic background radiation, such as the cosmic microwave background radiation and infrared background, and light provide a mechanism for the transmission of the Space-Time, Info-Energy, and Mass-Energy transformations in our dynamic world. The study of the 2-dimensional representations of background radiation will provide a vast spectrum of opportunities for a broad range of the natural sciences.

The Genome of the Universe

Time, space, energy and associated information are the fundamental characteristics of the Universe and every existing object and system. Nevertheless, the main secret of the Universe is the 'genome' of the Universe. It is kept as the time-associated latent energy that is coded and fixed by the latent information within the different representations of background radiation in the harmony of our dynamic world.

NASA published the following images of the Cosmic Infrared Background.

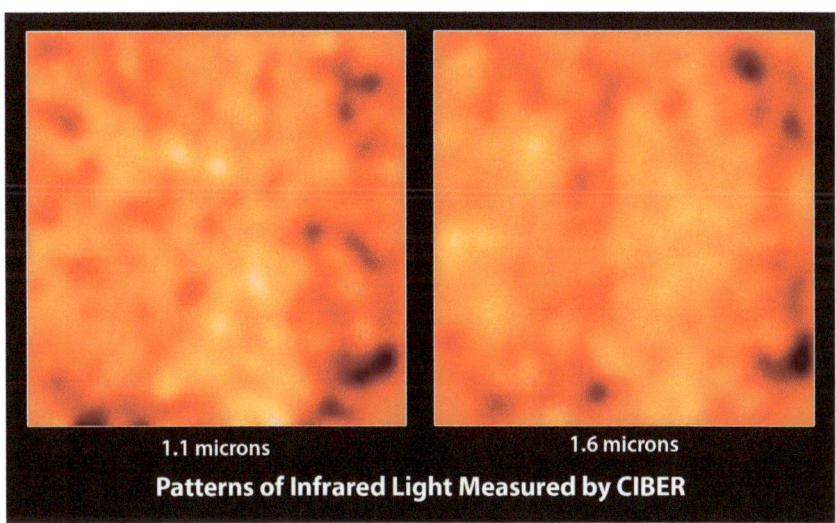

Image 8: 'Matching Patterns of Light' Instrument: CIBER Image credit: NASA/JPL-Caltech

'These images from the Cosmic Infrared Background Experiment, or CIBER, show large patches of the sky at two different infrared wavelengths (1.1 microns and 1.6 microns) after all known galaxies have been subtracted out and the images smoothed to enhance the large structures. CIBER sees similar patterns at different wavelengths, supporting the idea that the light patterns arise from the same source... 'Observations from NASA's Cosmic Infrared Background Experiment, or CIBER, have shown a surprising surplus of infrared light filling the Spaces between galaxies.' NASA/JPL-Caltech

We suppose that CIBER is looking at the Genome of the Universe.

The genome of the Universe is a complete set of the units which determine the fundamental characteristics of the Universe and the existing objects and systems. We can only guess what kind of information is kept as a genetic code of our Universe and every object and system, including stars, galaxies, and visually empty spaces, existing in the Universe like cells and functional systems of its body.

The dynamic body of the thermodynamic Universe integrates space, time, energy, and matter. Background radiation is similar to the vegetative nervous system in the body of the Universe - it regulates bodily functions of our dynamic world. Dynamic representations of background radiation, arranging the Universe latent Space-Time and Info-Energy structure, support, transport, and transmit the 'genome' of the Universe in our dynamic world.

There is no empty space in the Universe, and no isolated objects and systems exist in our dynamic world. Objects and systems, existing in the Universe, are tangled together by Space-Time, Info-Energy, and Mass-Energy imbalance and balancing transformations. The Universe contains 'the infinity' of Matrixes, tangled together in space and time of the Universe. Energies of background radiation and associated information support the latent Info-Energy of the

2-dimensional grid of every multimodal structure existing in the Universe. The complex space-time-info-energy structure of background radiation makes multimodal objects and systems the STIE-blocks (space-time-info-energy building blocks) of the Universe. Background radiation, carrying the 'genetic code' of our Universe, is the background structure of our dynamic world, containing its Space-Time, Mass-Energy, and Info-Energy characteristics. Dynamic representations of background radiation arrange the internal framework and Space-Time and Info-Energy skeleton of the multimodal Universe and objects and systems existing in our dynamic world.

Objects and systems, existing in the Universe, are tangled together by Space-Time and Info-Energy imbalance and balancing transformations. The frequency of the electromagnetic waves and their wavelength influence the 2-dimensional grids, changing objects and systems' properties through the energy transfer ruled by the Laws Space-Time, Mass-Energy, and Info-Energy Conservation, Reversibility, Limitation, Transformation, and Symmetry.

According to the Theory of Matrix, the genome of the Universe is a complete set of data establishing the fundamental characteristics of the Universe and the existing objects and systems. Principles of the genome decoding could include the laws of thermodynamics, the principle of balance, the speed of light in a vacuum that defines the exact proportion of space to time as the upper limit of our dynamic world, and Planck length, Planck time, and Planck energy, which define the exact lower limits of the existing objects and systems and the Universe as we measure and sense it. Background radiation and light carry inertial mass from the emitting system to the absorbing system. Different forms of background radiation and light propagate with the same speed - the speed of light in vacuum. They constitute the dynamic body of the thermodynamic Universe.

Anisotropies, or irregularities, of background radiation, such as the irregularities of the cosmic microwave background radiation and infrared background, are associated with the reorganisation of space, time, matter, energy, and information in the Universe. Besides, the irregularities of background radiation are associated with the generation of new multimodal objects and systems in the Universe and the incorporation of the objects and systems existing in our dynamic world.

Dark Matter And Dark Energy

The Universe is filled with the actual Info-Energy. We perceive the actual Info-Energy in different forms of energy and matter.

Image 9: 'The Clumping Behavior of Galaxies' Image credit: NASA/JPL-Caltech

'Normal matter, which is made up of atoms, is only a small percent of the total mass in our universe. Most of the matter in the universe is dark - that is, it does not emit or absorb any light...' NASA

Dark matter is 'non-luminous material which is postulated to exist in space and which could take either of two forms: weakly interacting particles (cold dark matter) or high-energy randomly moving particles created soon after the Big Bang (hot dark matter)'. (Oxford Dictionary) However, Dark energy is 'a theoretical form of energy postulated to act in opposition to gravity and to occupy the entire universe, accounting for most of the energy in it and causing its expansion to accelerate.' (Oxford Dictionary)

'Astronomers think that the expansion of the universe is regulated by both the force of gravity, which acts to slow it down, and mysterious dark energy, which pushes matter and space apart. In fact, dark energy is thought to be pushing the cosmos apart at faster and faster speeds, causing our universe's expansion to accelerate.

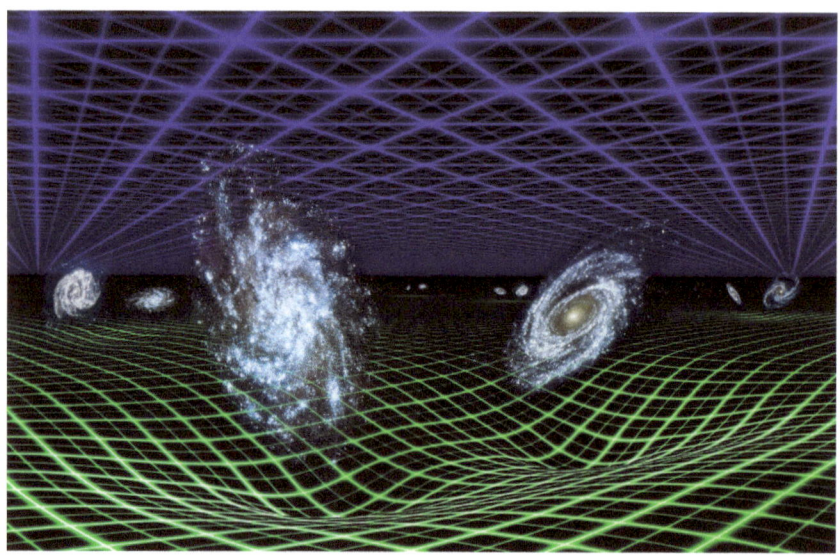

Image 10: 'Dark Energy and Gravity: Yin and Yang of the Universe (Artist's Concept)' Image credit: NASA/JPL-Caltech

'In this artist's conception, dark energy is represented by the purple grid above, and gravity by the green grid below. Gravity emanates from all matter in the universe, but its effects are localized and drop off quickly over large distances.

'New results from NASA's Galaxy Evolution Explorer and the Anglo-Australian Telescope atop Siding Spring Mountain in Australia confirm that dark energy is a smooth, uniform force that now dominates over the effects of gravity. The observations follow from careful measurements of the separations between pairs of galaxies (examples of such pairs are illustrated here). The results are one of the best confirmations of the nature of dark energy to date.' NASA

To clarify the ways of the Dark matter and Dark energy formation, we consider the relationship between gravitational and anti-gravitational processes in Space-Rising non-radiating systems.

Under the Theory of Matrix, the presence of Dark energy, Dark matter, and other forms of non-radiating matter indicate the anti-gravitational processes in the area.

Space and mass creation is prompted by the central areas of the Space-Rising non-radiating objects and systems. It coexists in a dynamic balance with space and mass degeneration and reduction prompted by their periphery. The generated space, energy, and matter, all decline throughout the system into the various forms of the degenerated space, non-radiating matter, and energy. They are detectable in the regions of inflation within the Universe.

We mentioned above that gravity is most intense at the centre of the Universe. Gravitational processes at the centre of the Universe prompt the local gravitational processes associated with space, kinetic energy, and matter generation throughout the Universe. The balancing transformations are reflected in gravitational acceleration. Gravity, gravitational acceleration, and generation of space, kinetic energy, and matter might be detected at the centres of the spacious and Space-Rising non-radiating systems.

The projection of the different representations of the 'o' time-point, existing in the SRMs' 2-dimensional time settings, into the volume of the Space-Rising non-radiating systems, for example, the toroidal systems, would give us the Space-Time regions with the qualities of gravity and generation of space, mass, and kinetic energy located within the projection at the centre of the system.

The Space-Time, Info-Energy, and Mass-Energy imbalance, reflected in gravity, can be responsible for the compensatory creation of the transmitting Black Holes on the periphery and at the centres of the large Space-Rising non-radiating systems. These transmitting Black Holes carry space, matter, and energy from the outer space into the central regions of their maternal non-radiating systems.

Antigravity is most intense in the peripheral regions of the Universe. Anti-gravitational processes on the periphery of the Universe prompt the local anti-gravitational processes associated with space, kinetic energy, and matter degeneration and reduction throughout the Universe. The balancing transformations are reflected in anti-gravitational deceleration. Antigravity, anti-gravitational deceleration, and degeneration and reduction of space, kinetic energy, and matter are prompted by the peripheral regions of the large spacious and Space-Rising non-radiating systems.

The Space-Time, Info-Energy, and Mass-Energy imbalance, reflected in antigravity, can be responsible for the compensatory creation of the transmitting Black Holes on the periphery and in the central regions of the large Space-Rising non-radiating systems. These transmitting Black Holes carry space, matter, and energy away from the maternal non-radiating systems (into the outer space).

Gravity at the centre of the Universe and antigravity in its peripheral regions bring about gravitational and anti-gravitational processes and balancing transformations within the system. Space, energy, and matter, generated under the influence of gravity, degenerate (being deprived of some physical qualities considered normal) through the system under the influence of antigravity creating degenerated (excessive and inflated) space, matter, and energy - Dark matter and Dark energy.

'Astronomers don't know why the hidden Black Holes would have larger halos of dark matter but are intrigued by the surprising finding and are investigating further.' NASA

Perhaps, we should pay attention to the Black Holes, balancing the anti-gravitational processes within the 2-dimensional Info-Energy grid in the peripheral areas of the large unbalanced Space-Rising and toroidal non-radiating objects and systems with an extensive collection of degenerated space, energy, and matter, including Dark matter and Dark energy.

If a Space-Rising non-radiating system undergoes the process of its natural development, the resultant force of its SRM reflects the influence of the non-radiating system's vector-force of resistance. It brings about the inflation of the Space of the Current Time. In the systems trending toward antigravity, 'normal' matter, energy, and space degenerate into the inflated space, degenerated non-radiating matter and energy, and pass into the reduction - the volumes, matter, and energy are being transformed into the 2-dimensional Space-Time supported by the latent, or potential, Info-Energy thus shifting the power from the centre to the periphery of the system, and finally transforming the Space-Rising non-radiating system into the toroidal system.

The gravitational and anti-gravitational horizon is the system's 2-dimensional Info-Energy grid, processing Space-Time and Info-Energy transformations. Dynamic representations of background radiation arrange, support, transport, and transmit the balancing Space-Time, Info-Energy, and Mass-Energy transformations, including those reflected in gravitational acceleration and anti-gravitational deceleration, through our dynamic world.

Image 11: 'Dark Matter 'Hairs' Around Earth - Close-up'
Image credit: NASA/JPL-Caltech

'This illustration shows Earth surrounded by filaments of dark matter called "hairs," which are proposed in a study in the Astrophysical Journal by Gary Prézeau of NASA's Jet Propulsion Laboratory, Pasadena, California.

'A hair is created when a stream of dark matter particles goes through the planet. According to simulations, the hair is densest at a point called the "root." When particles of a dark matter stream pass through the core of Earth, they form a hair whose root has a particle density about a billion times greater than average.

'The hairs in this illustration are not to scale. Simulations show that the roots of such hairs can be 600,000 miles (1 million kilometers) from Earth, while Earth's radius is only about 4,000 miles (6,400 kilometers).' NASA

Black Matter and Black Energy

Gary Prézeau of NASA's Jet Propulsion Laboratory, Pasadena, California, has proposed that the Earth 'surrounded by filaments of dark matter called "hairs".

We argue that the presence of non-radiating matter and energy indicates the anti-gravitational processes in the area. Gravitational processes reflect the generation of space, matter, and kinetic energy in the Universe. Anti-gravitational processes reflect the degeneration of space, matter, and kinetic energy in the Universe.

The generation of space, energy, and matter within the system under the influence of gravity, coexists in a dynamic balance with the degeneration of space, energy, and matter into different forms of non-radiating matter and energy prompted by antigravity.

Dynamic representations of background radiation arrange, support, transport, and transmit the gravitational and anti-gravitational processes through our dynamic world.

On the one hand, we observe and experience gravity on the surface of the Earth. The Space-Time, Info-Energy, and Mass-Energy imbalance, reflected in gravity, results in the forces providing balancing transformations reflected in gravitational acceleration.

Neither Mass-Energy nor the forces but the gravitational acceleration is the definite proof of gravity existence. The gravitational acceleration of the object's fall does not depend on the mass of the object. Mass and energy arrange the detectable physical basis for the Space-Time imbalance and balancing transformations.

The high energy massive radiating systems are not spherically symmetrical. The specific geometry of the Earth is reflected, for example, in gravitational acceleration on the surface of the Earth.

On the Earth, objects fall with a standard value of acceleration 9.80665 m/s². In the process of the gravitational acceleration, we can detect the relationship between space and time as the linear space per time squared.

Nevertheless, at different points on Earth, objects fall with an acceleration between 9.764 m/s² and 9.834 m/s² depending on altitude and latitude. The gravitational acceleration increases from about 9.780 m/s² at the equator to about 9.832 m/s² at the poles, because the relations between space and time are various at different points on the surface of the Earth (Figure 5) and the geometry of the planet is affected by its geometry in the 2-dimensional time settings.

The gravitational acceleration, reflecting the relationship between space and time, is the element that let us recognise the gravity and the degree of gravity, measured by acceleration.

On the other hand, we suppose that the central region of our planet, located within the projection of the different representations of the '0' time-point existing in the TRM 2-dimensional time settings (Figure 6), develops the Space-Time imbalance, such as the excessive space and time deficit, and the associated Info-Energy and Mass-Energy imbalance.

This fascinating Space-Time region is located within the projection at the centre of the Earth in our dynamic world. The Space-Time imbalance and the associated Mass-Energy and Info-Energy imbalance at the centre of the Earth are reflected in qualities of antigravity. We remember the observer who had reached the centre of the Earth and was stopped by the 'natural limit' - the impossibility of moving further to the centre and more profound.

The balancing transformations, reflected in anti-gravitational deceleration, enable space, energy, and matter degeneration and reduction. Probably, the elements of the degenerated matter and energy are not so far from us as we think.

We call the degenerated matter and energy, associated with anti-gravitational processes within the central regions of the high energy massive radiating systems, 'Black matter' and 'Black energy' to express the difference between the forms of degenerated - excessive, super-concentrated, deprived of some physical qualities considered normal space, matter, and energy at the centres of high energy massive radiating systems and the degenerated matter and energy within non-radiating systems (known as 'Dark matter' and 'Dark energy') according to the precisely opposite directions of the time and time-associated energy flow at the centres of these objects and systems (please compare Figures 3 and 4).

The difference in qualities of Black matter and Dark matter is comparable to the difference in qualities of the degenerated suppressed amorphous matter with the dissociated atomic bonds at the centres of the high energy massive radiating systems and the hard vacuum of outer space.

Space, energy, and matter, generated under the influences of the system's vector-force of resistance prompted by the peripheral regions of the high energy massive radiating system, degenerate throughout the system into different forms of non-radiating matter and energy under the influence of the Matrix grid vector-force of pressure prompted by the central regions of the system, such as the Sun, the Earth, the globular star clusters and galaxies, and other high energy massive radiating systems.

Accordingly, at the centre of the Earth, the influence of the Matrix grid vector-force of pressure is associated with the Matrix tendency to transform the Space of the Current Time into the system's 2-dimensional Space-Time, along with the transformations of the actual Info-Energy, acting in the volume of our planet as kinetic energy and mass, into the potential Info-Energy of the 2-dimensional grid arranged by background radiation. These transformations are reflected in antigravity propagation. The dominant influence of the 2-dimensional grid vector-force of pressure, supported by background radiation, provides the degeneration and reduction of space, mass, and kinetic energy reflected in the anti-gravitational processes in the region.

Dynamic representations of background radiation arrange, support, transport, and transmit the gravitational and anti-gravitational processes through our dynamic world. If balancing transformations are virtually ineffective, the transmitting Black Hole might be formed at the centre of the high energy massive radiating system or in its peripheral regions.

Reverse of the Time and Energy Flow

We mentioned above that there is no empty space in the Universe. The 'outer space' is always another object or a system, for example, a non-radiating system, mainly conducting space and mass degeneration and reduction, or it is a high energy massive radiating system tangled by gravity such as our planet and the Sun. Objects and systems are tangled together with the Space-Time and Info-Energy imbalance and balancing transformations. Background radiation, arranging the 2-dimensional grid of the objects and systems' multimodal structures, provide the mechanism for the Space-Time, Info-Energy, and Mass-Energy transformations and their transmission in our dynamic world.

Planck length, Planck time, and Planck energy, as necessary deviations from the '0' Space-Time-Energy point, define the minimal conditions of an object or a system's existence. Accordingly, they are the factors influencing the reverse of the Space-Time and Info-Energy flow in the multimodal structures of the existing objects and systems. The speed of light in a vacuum is the upper limit activating the system's reverse. If an object or a system has reached these limits, the direction of its Space-Time and Info-Energy 'flow' is being reversed by the Matrix forces. The reverse of the time and energy 'flow' within the system's multimodal structure will conduct the reverse of the Space-Time and Info-Energy flow within the system in our dynamic world.

How would the reverse of the Space-Time and Info-Energy flow occur in our dynamic world?

The process would have some similarity with the magnetic field reversal when the system's polar magnetic fields go to zero and then emerge again with the opposite polarity.

Please consider the reversal effects, related to the quantum space, quantum time, and quantum energy, which are associated with the Space of the Current Time at the Matrix centre.

Quantum energy as a discrete packet of energy can be absorbed, or transformed into the form of the latent Info-Energy of the Matrix grid. The discrete packet of energy can be released, or transformed into the form of actual Info-Energy acting as a quantum in our dynamic world. If an object or a system has reached the lower limits actuating a reverse (Planck length, Planck time, and Planck energy), it is being reversed by the Matrix forces. Accordingly, a particle or energy appears or disappears in a vacuum as a result of the reverse of the Space-Time and Info-Energy flow within the multimodal system and the associated balancing Space-Time, Info-Energy, and Mass-Energy transformations.

Changes of the frequency of electromagnetic waves and their wavelength affect the 2-dimensional grid, changing Space-Time, Info-Energy, and Mass-Energy properties of the particle through the energy transfer. Dynamics preserve the laws of Space-Time, Mass-Energy, and Info-Energy Reversibility, Limitation, Conservation, Transformation, and Symmetry.

In the macro-world, high energy massive radiating systems, such as radiating galaxies and star clusters, stars and planets, and Space-Rising non-radiating systems, such as the self-rising vacuum and systems holding Black Holes, display the Space-Time, Mass-Energy, and Info-Energy imbalance and balancing transformations. The reverse and reorganization of matter and energy create new radiating and non-radiating objects and systems in our dynamic world.

The system's reversibility is determined by the Space-Time, Mass-Energy, and Info-Energy resources of the system. If balancing transformations are virtually ineffective and the system reaches the upper or lower limits activating reverse, the Matrix forces will reverse the system's multimodal structure within limits ruled by the Laws of Space-Time, Info-Energy, and Mass-Energy Reversibility, Limitation, Conservation, Transformation, and Symmetry.

The resultant vector-force of the system's multimodal structure will reverse the direction of the Space-Time and Info-Energy flow in the Spaces of Time. The direction of Arrows of Time provides us with an opportunity to detect the possible reverse of the time and energy flow within the multimodal systems.

Following the reverse of the Space-Time and Info-Energy flow, the resultant vector-force applies the balancing transformations to establish a new balance and symmetry of the multimodal Space-Time and Info-Energy structure of the system.

The reverse of the Space-Time and Info-Energy flow within the system's multimodal structure is reflected in the reorganization of the volumes, matter, and energy of the system in our dynamic world.

The continuing space and mass degeneration, prompted by the centre of a high energy massive radiating system, such as a star or a radiating galaxy tangled by gravity, might dominate the processes of the space and mass generation prompted by the peripheral areas of this system. It would result in the rising Space-Time and Info-Energy imbalance under the influence of the 2-dimensional grid vector-force of pressure developing the system's time, Potential Space, and the associated potential Info-Energy and shaping the toroidal form of the system in the multimodal Space-Time. This Space-Time and Info-Energy imbalance would be reflected in the degeneration of space, energy, and mass into different forms of the degenerated matter and energy, such as the Black matter and Black energy, their reduction, and accordingly, limitation of the actual space, mass, and kinetic energy resources of the multimodal system, along with the growing antigravity and anti-gravitational acceleration prompted by the centre of the system.

The Space-Time and Info-Energy imbalance and the associated anti-gravitational processes can damage the system's centre and create the central Black Hole that ejects the Potential Space and the space-associated potential Info-Energy from the system. The compensatory transmitting Black Holes might be created in the peripheral areas of the system.

The high energy massive radiating systems, reaching the limits actuating a reverse, will be reversed by the Matrix forces into the Space-Rising non-radiating systems (Figure 10).

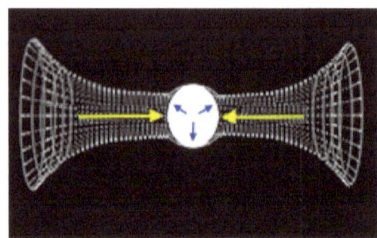

Figure 10: TRM Reverse

The reverse of the Space-Time and Info-Energy flow within the system's multimodal Space-Time and Info-Energy structure initiates the development of the deflation, gravitational acceleration, and creation of volumes, masses, kinetic energy and other energy representations at the system's centre. The high energy massive radiating system will be reversed into the Space-Rising non-radiating system. Accordingly, the system's TRM is being reversed into the system's SRM (Figure 10).

The reversed Space-Time and Info-Energy flow within the central Black Hole will transmit volumes, masses, and kinetic energy, along with the Potential Space and the space-associated potential Info-Energy from outer space into the system.

Alternatively, the continuing space and mass generation, prompted by the periphery of a high energy massive radiating system, such as our Sun, might dominate the space and mass degeneration prompted by its central areas.

It would result in the rising Space-Time and Info-Energy imbalance under the influence of the system's vector-force of resistance developing the system's Space of the Current Time, along with the associated actual Info-Energy represented in mass and kinetic energy acting in the volume of the system in the period of time now occurring.

The Space-Time and Info-Energy imbalance and the dominant influence of the system's vector-force of resistance would lead to the concentration and consolidation of space and matter reflected in the intense gravity prompted by the peripheral regions of the system, along with the limitation of the 2-dimensional Space-Time, potential energy resources, and creation of the transmitting Black Holes within the system's 2-dimensional grid. The transmitting Black Holes within the system's 2-dimensional grid will transmit the system's Potential Space and the associated potential energy from the system into outer space.

The excessive actual space and masses may damage the system's centre and create the central Black Hole. The central Black Hole will transmit volumes, masses, and kinetic energy from the system into outer space.

If the system has reached the limits actuating the reverse, it is being reversed by the Matrix forces. The limitation of the system's potential Space-Time and Info-Energy resources will reverse the Space-Time and Info-Energy flow within the system and develop the inflation and reduction of the system's volumes, masses, and kinetic energy. The process would be reflected in anti-gravitational deceleration prompted by the peripheral regions of the system.

The reverse of the Space-Time and Info-Energy flow within the system will reverse the Space-Time and Info-Energy flow in the system's Black Holes to regain the potential resources of the system. Simultaneously, the central Black Hole will transmit volumes, masses, and kinetic energy from the outer space into the system's centre.

The following deflation and generation of volumes, masses, and kinetic energy, reflected in gravitational acceleration prompted by the system's centre, would reverse the radiating system into the Space-Rising non-radiating system. Accordingly, the system's TRM will be reversed into the SRM (Figure 10).

The large Space-Rising non-radiating systems, reaching the limits of the system's existence, will be reversed into the high energy massive radiating systems (Figure 11).

The continuing space and mass generation, prompted by the centre of a spacious non-radiating system, might dominate the space and mass degeneration, prompted by the system's peripheral areas. It would result in the rising Space-Time and Info-Energy imbalance under the influence of the system's vector-force of resistance developing the system's Space of the Current Time and associated actual Info-Energy represented in mass and kinetic energy acting in the volume of the system in the period of time now occurring. The Space-Time and Info-Energy imbalance is reflected in the intense gravity prompted by the centre of the system. The excessive actual space and masses can damage the system's centre and create the central Black Hole.

The process coexists with the rising deficit of the Potential Space, time, and the associated potential Info-Energy resources, damage of the 2-dimensional grid, and creation of the transmitting Black Holes within the system's 2-dimensional grid.

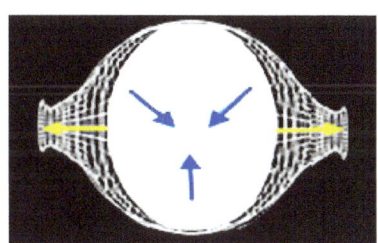

Figure 11: SRM Reverse

The following limitation of the potential Space-Time and Info-Energy resources will reverse the Space-Time and Info-Energy flow within the system's Spaces of Time and the system's Black Holes. It will bring about the generation of spaces and masses, along with the gravitational processes prompted by the periphery of the system and the development of inflation, anti-gravitational deceleration, and reduction of volumes, masses, kinetic energy, and other energy representations prompted by the system's centre. The non-radiating system will be reversed into the high energy massive radiating system. Accordingly, the system's SRM will be reversed into the TRM (Figure 11).

Alternatively, the continuing space, kinetic energy, and mass degeneration and reduction, along with anti-gravitational deceleration prompted by the periphery of the non-radiating system, might dominate the processes of the space, kinetic energy, and mass generation reflected in gravitational acceleration and prompted by the system's centre. It would lead to the development of the toroidal form of the system in the 2-dimensional Space-Time and the intense antigravity prompted by the periphery of the system. The deficit of the actual space, masses, and kinetic energy could damage the centre of the system and form the central Black Hole.

The following limitation of the actual Space-Time and Info-Energy resources of the system will reverse the direction of the Space-Time and Info-Energy flow (Figure 11) and initiate deflation, gravitational acceleration, and generation of volumes, masses, and kinetic energy prompted by the peripheral regions of the system. The reverse of the Space-Time and Info-Energy flow within the system will reverse the Space-Time and Info-Energy flow in the system's Black Hole. The central Black Hole will transmit the 2-dimensional Space-Time and associated potential Info-Energy, along with the newly created space and matter from the system's centre into the outer space.

The Space-Rising non-radiating system will be reversed into the high energy massive radiating system. Accordingly, the system's SRM will be reversed into the TRM (Figure 11).

The reverse of the direction of the Space-Time and Info-Energy flow will change the properties of the system. The reversed non-radiating systems, passing the period of their development into the high energy massive radiating systems, create supernovae in our dynamic world.

The ring galaxies with external accretion and cohesion of matter under the influence of gravitation, prompted by the periphery of the galaxy, would be an excellent example of the systems developing the surface orientated transmitting Black Holes.

Image 12: 'Ring of Stellar Fire' Image credit: NASA/JPL-Caltech

'This image from NASA's Spitzer Space Telescope, taken in infrared light, shows where the action is taking place in galaxy NGC 1291.

'The outer ring, colored red in this view, is filled with new stars that are igniting and heating up dust that glows with infrared light. The stars in the central area produce shorter-wavelength infrared light than that seen in the ring, and are colored blue. This central area is where older stars live, having long ago gobbled up the available gas supply, or fuel, for making new stars. The galaxy is about 12 billion years old and is located in the Eridanus constellation. It is known as a barred galaxy because a central bar of stars (which looks like a blue "S" in this view) dominates its center.

'When galaxies are young and gas-rich, stellar bars drive gas toward the center, feeding star formation. Over time, as the star-making fuel runs out, the central regions become quiescent and star-formation activity shifts to the outskirts of a galaxy. There, spiral density waves and resonances induced by the central bar help convert gas to stars. The outer ring, seen here in red, is one such resonance location, where gas has been trapped and ignited into a star-forming frenzy.' NASA

The ring galaxy (Image 12) with external accretion and cohesion of matter under the influence of gravitation could be one of two types. It can be formed by the radiating Black Hole at the centre of the galaxy and possible supernova explosion. Alternatively, the ring galaxy with external accretion and cohesion of matter under the influence of gravitation, prompted by the periphery of the galaxy, could be a reversed toroidal system being on the intermediate stage of the development into the globular cluster.

Should we travel through the ring galaxy towards the galactic centre, we would experience the deficit of space and the excessive time with time dilation and gravity on the periphery of the galaxies and the resistance of the space on the way closer to the centre as a result of the excessive concentrated space with space dilation and the deficit of time, and enforced anti-gravitational deceleration prompted by the galactic centre. It would resemble climbing a steep hill in low gear.

In the General Theory of Relativity, Albert Einstein described gravity as a consequence of the curvature of Space-Time. In space, we have 'hills' with antigravity, and we have 'valleys' with gravity. And we have the 'flat land' where gravity and antigravity coexist in a dynamic balance.

Our young Universe, rising in size, continues progressing as the Space-Rising non-radiating system. If our Universe stretches to the state of the 2-dimensional toroid and reaches the limits actuating the reverse, it will be reversed by the Matrix forces.

The reverse of the direction of the Space-Time and Info-Energy flow will initiate deflation, gravitational acceleration, and generation of volumes, masses, kinetic energy, and other energy representations prompted by the peripheral regions of the Universe.

Accordingly, the Universe would be reversed into the high energy massive radiating system and could create the supernova in the associated modality of the Grand Universe.

The Big Bang theory as a cosmological model of the Universe is struggling with the starting point of the Universe existence in time, space, mass, and energy. It also has difficulties with the Mass-Energy Conservation Law. It does not have to be that way. Nothing is coming from nothing and going to nothing.

We must not put the principle of conservation of energy aside and ignore the basics of the quantum theory.

The '0' Space-Time and '0' energy point as the start or the end of the Universe will never be found or proven because they never exist. They are not achievable in our dynamic world of matter and related exclusively to the human perception of time as a point 'now'.

We mentioned above that Planck length, Planck time, and Planck energy reflect the minimal Space-Time and Info-Energy imbalance in the existing objects and systems as the necessary deviations from the '0' Space-Time point and '0' energy point.

The speed of light in vacuum and the Planck units, based on the calculations using the speed of light in vacuum as the fundamentals proven by experiments - Planck length, Planck time, and Planck energy, provide the upper and lower limits for the existence of the Universe as we measure and sense it. If a system has reached the limits actuating the reverse, the Space-Time and Info-Energy flow within the system is being reversed following the Laws of Space-Time, Mass-Energy, and Info-Energy Conservation, Reversibility, Limitation, Transformation, and Symmetry.

Rotation

Space sometimes is looking flat and time moves the world. Masses, associated with the visible space, are rotated by the centrifugal force inside the temporal Matrixes, around their mass centres, at '0' time-points of our understanding, like the surface of water rotating in a bucket. Space-Time flux changes properties and builds channels, path shorting and bringing forth spaces close together or further apart.

Rotation of high energy massive radiating systems and Space-Rising non-radiating systems about their Space-Time axis is detectable in the first time dimension (Figures 1, 2). It affects the associated rotation of these systems about their axis of rotation in our dynamic world.

The system's Space-Time imbalance and the associated Info-Energy and Mass-Energy imbalance, detectable in the second time dimension (Figures 3, 4) and 2-dimensional time settings, are reflected in their geometry that affects the rotation of these systems about their Space-Time axis. It impacts their associated rotation about the axis of rotation in our dynamic world.

The system's Space-Time imbalance and associated Info-Energy and Mass-Energy imbalance activate the two opposite contact vector-forces - the Info-Energy grid vector-force of pressure and the system's vector-force of resistance, which are responsible for the rotation of the unbalanced systems, such as planets, stars, and galaxies, about their axis of rotation. Pressure across the surface of contact leads to the rotation of multimodal systems about their Space-Time axis in the Matrix. It impacts their associated rotation about the axis of rotation in our dynamic world with the gradual shift in the orientation of the axis of rotation, or the precession.

Please consider an example. Schrödinger equation for a particle, encountering a rectangular potential energy barrier, describes the conditions for a multimodal particle at the centre of its Matrix, at the 'o' time-point - the point 'now'. Under the Theory of Matrix, a multimodal particle, which impinges on the barrier from one side, demonstrates Space-Time and Info-Energy imbalance, similar to other high energy radiating objects and systems, such as planets and stars. The Space-Time imbalance and the associated Info-Energy imbalance are reflected in gravity or antigravity - in case of a microscopic hole.

The rectangular barrier is associated with the 2-dimensional rectangular Info-Energy grid arranged by the 2-dimensional representation of electromagnetic radiation. The change of the frequency of the electromagnetic waves would be an elegant, practical, real-life decision of the problems associated with the rectangular barrier. This influence on the Space-Time and Info-Energy imbalance of the blocked particle will change the properties of its 2-dimensional rectangular Info-Energy grid, and accordingly, change the particle behaviour, including such qualities as rotation and gravity.

The system's vector-forces activate balancing transformations within the systems.

Balancing transformations, reflected in the reorganisation of space, time, matter, and energy, have a secondary effect on the rotation of the unbalanced radiating and non-radiating objects and systems.

Four different types of rotation depend on the type of the Space-Time and Info-Energy imbalance and the dominant vector-force acting in the multimodal Space-Time and Info-Energy structure of the system.

The Space-Time imbalance and the associated Info-Energy and Mass-Energy imbalance within the high energy massive radiating systems and reversed non-radiating systems are reflected in gravity prompted by the peripheral areas of these systems and antigravity prompted by their central regions. Accordingly, the rotation of these systems might be determined by the Info-Energy grid vector-force of pressure prompted by the centre of the system or the system's vector-force of resistance prompted by the system's periphery.

The Space-Time imbalance and the associated Info-Energy and Mass-Energy imbalance within large spacious non-radiating systems and reversed high energy massive radiating systems are reflected in antigravity prompted by the peripheral areas of these systems and gravity prompted by their centres. Accordingly, the rotation of these systems might be determined by the Info-Energy grid vector-force of pressure prompted by the periphery of the system or the system's vector-force of resistance prompted by the system's centre.

The influence of the Matrix forces and rotation of the unbalanced objects and systems about their Space-Time axis affect and direct the associated rotation of the 2-dimensional grid that is the external framework of the system in its Matrix and, simultaneously, the dynamic internal skeleton of this system in our dynamic world.

The 2-dimensional representations of background radiation, currently known as cosmic background radiation, and light arrange and support the 2-dimensional grid of the existing objects and systems. The relative specific rotation of the Matrix grid is reflected in the rotation of background radiation and light that must be detectable.

We mentioned above that unbalanced systems might develop the transmitting Black Holes. The Black Holes connect the central points of symmetry of the associated Matrixes and the shared 2-dimensional Space-Time areas of the Matrixes connection. Please see my book 'Black Holes and Supernovas' for details. The rotation of the unbalanced system impacts the rotation of the associated Black Hole, the flow of matter, volumes, and energy, which this Black Hole delivers to the system, and the rotation of the Space-Time region associated with the unbalanced system through the Black Hole.

The reverse of an unbalanced multimodal system affects the specific rotation of this system and the associated rotation of the background radiation. The newly established Space-Time and Info-Energy imbalance changes the direction of rotation to the opposite in the Matrix and impacts the rotation of the reversed system in our dynamic world.

The rotation of the reversed system about its axis of rotation would reflect a new direction of its Space-Time axis and, accordingly, a new axis of rotation in our dynamic world. It would impact on the gradual shift in the orientation of the system's axis of rotation.

The reverse of an unbalanced multimodal system, holding the Black Hole, affects the rotation of the Black Hole, the flow of matter, volumes, and energy, which this Black Hole transmits, and rotation of the associated Space-Time region, such as a star or a visually empty region of space, connected to the reversed system through the Black Hole.

The reversed system ejects the potential space and associated latent Info-Energy, along with the created masses and kinetic energy, which continue their rotation while being ejected from the central areas of the reversed system through the Black Hole into outer space.

Our young Universe, rising in size, is in the process of its natural development. It continues progressing as the Space-Rising non-radiating system utilising volumes, energy, and associated information received through the giant central Black Hole from the radiating modality of the Grand Universe. Accordingly, our Universe is greatly influenced by the radiating modality of the Grand Universe and anti-gravitational processes within its central areas, its rotation about the Space-Time axis, and rotation of background radiation supporting the 2-dimensional grid of the system's Matrix. The influence of the radiating modality of the Grand Universe makes an impact on the Universe axis of rotation and its precession in our dynamic world.

Main Principles

In this chapter, we reaffirm some principles of the Theory of Matrix, which have been discussed in the book. These principles are related to the natural phenomena and applicable to the Universe, its modalities, and the existing objects and systems. Taking into consideration the human perception of time and removing constraints from creative thought, we introduce new concepts of space and time concerning the multimodal Space-Time and the principle of Space-Time equivalence. Drawing inferences and obtaining conclusions from the fundamental laws of nature by a sequence of logical steps and deduction from the fundamentals proven by experiments, we can overcome the limitations of time-perception.

Principles of the Theory of Matrix apply to every existing object and every system of the objects, including visually 'empty' spaces, quanta, subatomic particles and holes, systems holding Black Holes and creating supernovae, other objects and systems existing in the Universe.

1. Neither empty space nor empty time exists in the Universe.

The space of the Universe is formed by and filled with energy.

We perceive this energy in different forms, such as masses, kinetic energy, and other energy representations acting in the volumes of the objects and systems existing in the Universe in the period of time now occurring. The potential energy and associated information build and fill the time of the Universe. Following our perception of time, we cannot directly sense the time-associated energy of the Universe, but we can detect the influence of the potential energy as and when we experience gravity. The latent, or potential, Info-Energy is the predominant form of energy in the Universe.

The latent, or potential, energy of the objects and systems, existing in the Universe, is coded and fixed by the latent, or potential, information within the different representations of background radiation. Background radiation, arranging the potential Info-Energy into the 2-dimensional framework within the multimodal structures of the objects and systems, builds simultaneously the internal framework and Space-Time and Info-Energy skeleton for every object and system existing in our dynamic world.

Background radiation and light constitute the background structure of our dynamic world.

2. Multimodal objects and systems of the multimodal Universe

Every existing single object and every system of the objects, including visually 'empty' spaces, subatomic particles and holes, quanta, humans, systems holding Black Holes, other objects and systems, and our Universe, are the multimodal objects and systems carrying the properties of various Space-Time modalities. The object or the system's characteristics are different in different modalities of Space-Time.

The multimodal Space-Time and Info-Energy structure of an object or a system we call the object or the system's Matrix for short. Accordingly, the Matrixes are the multimodal Space-Time and Info-Energy structures of the objects and systems with qualities of mass, volume, energy, and time of existence.

The 1-dimensional Space-Time, supported by energy, is an undifferentiated existence of space, time, and energy underlying our dynamic world. It is built by and filled with the one quality Info-Energy of infinite duration, presenting our dynamic world with the ultimate dynamics.

The object or the system's fundamental 2-dimensional Space-Time is built by and filled with the potential Info-Energy of the object or the system. The 2-dimensional time forms the object or the system's Spaces of Time. The 2-dimensional potential space underlies the volume of the object or the system.

The 3-dimensional Space-Time, or the volume of an object or a system, is built by and filled with the actual Info-Energy represented in mass, kinetic energy and other energy representations acting in the volume of the object or the system in the period of time now occurring.

3. Building blocks (STIE-blocks) of the Universe

The multimodal Space-Time and Info-Energy structures of the existing objects and systems are tangled together by the Space-Time, Info-Energy, and Mass-Energy imbalance and balancing transformations transmitted by background radiation. Dynamic representations of background radiation make the Matrixes the 'Space-Time-Info-Energy' building blocks (STIE-blocks) of the Universe.

4. Space-Time Equivalence

The universal proportionality exists between equivalent amounts of energy and mass,

$E = mc^2$,

where E is energy and m is mass.

The universal proportionality exists between equivalent amounts of time and space,

$Et^2 = [c]^2 ml^2$,

where E is energy, m is mass, l^2 is the 2-dimensional space, t^2 is the 2-dimensional time, and $[c]^2$ is the Coefficient of Transformation.

5. The Coefficient of Transformation $[c]^2$

The Coefficient of Transformation $[c]^2$ equals the squared numerical value of the speed of light in a vacuum.

The Coefficient of Transformation $[c]^2$ applies to the decisions associated with the Space-Time, Info-Energy, and Mass-Energy transformations. Dynamics preserve the Laws of Space-Time, Info-Energy, and Mass-Energy Conservation, Reversibility, Transformation, Limitation, and Symmetry.

The value of the speed of light in a vacuum (c) is 299 792 458 m s-1 (meter per second). It is the Fundamental Physical Constant. Under CODATA, the velocity of light measurements carried out by Michelson and his associates over the period 1924 to 1935. (NIST. Fundamental Physical Constants)

6. Relativity of motion

The concept of relativity and Einstein's Special Theory of Relativity state that all motion is relative and the velocity of light in a vacuum has a constant value which nothing can exceed.

We add that the relative difference between our perception of space and time makes this world dynamic for us - relations between space and time we perceive as speed and acceleration.

The speed of light (and background radiation) in a vacuum as the fundamental relation between 1-dimensional space and 1-dimensional time defines the exact proportion of space to time as the upper limit of the existing objects and systems and the Universe as we measure and sense it. Planck units, such as Planck length, Planck time, and Planck energy based on the calculations using the speed of light in vacuum as the fundamentals proven by experiments, provide the lower limits of the existing objects and systems and the Universe as we measure and sense it.

7. The Space-Time Coefficient [c]

All dynamics in our dynamic world is associated with the difference in our perception of space and time - the relation between 1-dimensional space and 1-dimensional time we perceive as speed. The numerical value of the velocity of light in a vacuum is the Space-Time Coefficient that defines the relation between 1-dimensional space and 1-dimensional time in a vacuum.

The Space-Time Coefficient [c] equals the numerical value of the speed of electromagnetic waves propagation in a vacuum.

The Space-Time Coefficient [c] applies to the decisions associated with the transmission of the Space-Time, Mass-Energy, and Info-Energy transformations, including those reflected in gravity and antigravity propagation in our dynamic world.

8. Relativity of time, space, mass, energy, and information

Time, space, mass, energy, and information do not exist independent of objects and systems of the objects. They are the properties of the existing objects and systems. Time, space, mass, energy, and information are to be defined in relation to a frame of reference.

Space-Time is to be defined in relation to a frame of reference. Objects and systems of limited volumes are limited in their time of existence and, accordingly, limited in Space-Time. Objects and systems of the limited time of existence are limited in Space-Time. We specify the relation between time and space. Time is relative to space, changing, and transforming into space, and space is relative to time, changing, and transforming into time.

Info-Energy is to be defined in relation to a frame of reference. Mass-Energy is 'the mass of a body regarded as energy, according to the laws of relativity' (Oxford Dictionary). Objects and systems of limited mass are limited in their energy, and objects and systems of limited energy are limited in their mass. Energy is relative to masses, changing, and transforming into masses, and masses are relative to energy, changing, and transforming into energy.

Objects and systems of limited mass and energy are limited in Space-Time.

Objects and systems of limited volumes and time of existence are limited in their Info-Energy.

9. Relation of time, space, mass, energy, and information to the observer

The time, space, mass, energy, and information of the existing objects and systems are relative to the observer. You remember the Schrödinger's cat. Only the observer could say the cat in the box is either alive or dead. Alternatively, the cat is not in the box in the first place.

The observer observes an event and measures it, but the normal limits of human perception of the event and the environment restrict his ability to observe the existing object or the system. The observer's observation abilities are limited in quantity (5) and quality of senses. We, humans, observe the world and the existing objects and systems using our senses, bodies, and memories. We use the receptors of our bodies with their limited ability to respond to the external stimuli, and it takes time to transmit the signal through the series of synapses, compare it with the preserved data in several layers of memories, and produce the sensible response.

We reproduce our limited diapason of senses in new technologies extending the quality of our perceptual abilities. Nevertheless, new technologies reflect our limited quantity of senses.

Accordingly, the world is relative to the observer and not defined.

10. The Systems Theory and the Principle of Balance in Cosmology

Space, time, mass, energy, and information of every object are integrated into space, time, mass, energy, and information of the system, similar to the water molecules in the ocean or points of space included in the set of points having some specified structure in mathematics.

Space, time, mass, energy, and information of an object in the system is entirely negligible relative to space, time, mass, energy, and information of the system and reflected in the resultant space, time, mass, energy, and information of the system. Similarly, water molecules are bonded together in the ocean, thus achieving a new quality.

Forces between objects in a system are integrated and reflected in the resultant force of the system that is built by these objects. Forces between objects in the system are entirely negligible relative to the resultant force of the system built by these objects.

Space, time, mass, energy, and information of the existing objects and systems are integrated and reflected in the resultant space, time, mass, energy, and information of the Universe. Forces between the existing objects and systems are integrated and reflected in the resultant force of the thermodynamic Universe.

The Space-Time, Info-Energy, and forces of the multiple Universe modalities, such as the thermodynamic Universe and other Universes, Multiverse, stages of the Universe development, and other modalities of the Grand Universe, are finally integrated into the Space-Time, Info-Energy, and force of the Grand Universe.

The Grand Universe includes and balances all modalities of the Universe, and its 'o' Space-Time point, as the point 'here' and 'now', is the theoretical space-time-mass-energy null-point. It is the 'o world point', or 'a space-time-null-point', mentioned in Special Relativity by H. Minkowski, A. Einstein, and H. Weyl in relation to the mathematical model of the Light Cone.

Actualities of the various existing modalities of the Universe, their potentialities, other properties, and numerous opposing forces are balanced through the theoretical 'o' space-time-mass-energy point and reflected in the resultant force of the Grand Universe. The resultant force of the Grand Universe equals zero.

The characteristics of different modalities of the Universe and various stages of the Universe development in complex combinations co-exist in the real multimodal world in the state of equilibrium, yielding a net-zero and providing us with the theoretical 'o' Space-Time-Info-Energy point - the world point. The 'o' point is the point of the absolute balance - the total equilibrium that can be applied to the Grand Universe and fails our dynamic world.

The multimodal Universe exists in the state of equilibrium that means the qualities cancel out, yielding a net quality of zero, thus making the Grand Universe neutral.

Accordingly, the multimodal Universe was not born. It has no start of the existence in space, time, mass, energy, and information. It has no end but modalities. This theoretically possible scenario could be applied only to the Great Multimodal Universe. A balanced, isolated object or an isolated system in the state of equilibrium would cease to exist.

Natural Laws of the Theory of Matrix

The Natural Laws of the Theory of Matrix include the Laws of Space-Time, Info-Energy, and Mass-Energy Reversibility, Conservation, Transformation, Limitation, and Symmetry. They reflect the natural phenomena and apply to the objects and systems existing in the Universe and the Universe modalities.

The Laws of Balance and Symmetry
Every existing multimodal object and every system of the objects display a tendency to obtain and retain the Space-Time, Info-Energy, and Mass-Energy balance and symmetry through the Space-Time axis and 'o' Space-Time point at the centre of the object or the system's multimodal Space-Time and Info-Energy structure. The Laws of Symmetry is followed on from Newton's first law.

Isaac Newton defined inertia as his first law. In the modern physics, inertia is understood as 'a property of matter by which it continues in its existing state of rest or uniform motion in a straight line, unless that state is changed by an external force' (Oxford dictionary).

Albert Einstein indicated that inertia is not a fundamental property of matter, but a property of energy.

To sum up, inertia is a property of matter and energy to remain unchanged or resist changes.

The Limitation Laws
Objects and systems of the objects, limited in space, are limited in time. The objects and systems, limited in time, are

limited in space. The multimodal structures of these objects and systems are limited in Space-Time.

The objects and systems of the objects of limited mass, energy, and associated information are limited in space and time. The objects and systems, limited in space and time, are limited in mass, energy, and associated information. The multimodal structures of the objects and systems of limited mass, energy, and associated information are limited in Space-Time.

The speed of light, Planck length, Planck time, and Planck energy provide the upper and lower limits for an object or a system's existence.

An object or a system, which is smaller than the wavelengths supporting the 2-dimensional grid of its multimodal Space-Time and Info-Energy structure, does not exist in Space-Time.

The Laws of Space-Time, Info-Energy, and Mass-Energy Limitation are ruled by the Laws of Space-Time, Info-Energy, and Mass-Energy Conservation, Transformations, Reversibility, and Symmetry. The Space-Time, Info-Energy, and Mass-Energy Limitation Laws apply to every existing object and every system.

The Laws of Space-Time, Info-Energy, and Mass-Energy Conservation

'By laying down the relativity postulate from the outset, sufficient means have been created for deducing henceforth the complete series of Laws of Mechanics from the principle of conservation of energy (and statements concerning the form of the energy) alone.' [Minkowski Hermann, The Fundamental Equations for Electromagnetic Processes in Moving Bodies. Appendix (1908)].

Regarding the above statement and the principle of conservation of energy of Émilie du Châtelet, we introduce the Laws of Space-Time, Info-Energy, and Mass-Energy Conservation.

1. The Law of Mass-Energy Conservation

Albert Einstein has specified the relation between mass and energy of the system as follows, 'a body of mass m is to be regarded as a store of energy of magnitude mc². ' [Einstein Albert, A Brief Outline of the Development of the Theory of Relativity (1921)]

The universal proportionality exists between equivalent amounts of energy and mass,

$E = mc^2$,

where E is energy and m is mass.

The total Mass-Energy of an existing system equals the total Mass-Energy that the system possesses for the total period of its existence, including its past, present, and future. The mass of the system may change, although the total Mass-Energy of the system remains constant.

The total Mass-Energy of an existing system is conserved in the system's multimodal Space-Time and Info-Energy structure.

2. The Law of Info-Energy Conservation

Mass-Energy is 'the mass of a body regarded as energy, according to the laws of relativity' (Oxford Dictionary). We, respecting the laws of relativity, use the term 'Mass-Energy' as interchangeable with the term 'Info-Energy' of a system.

The system's total energy and associated information equal the total Info-Energy that the system possesses for the total period of its existence, including its past, present, and future. The actual Info-Energy of the system may change, although the total Info-Energy of the system remains constant. The latent Info-Energy of the system may change, although the total Info-Energy of the system remains constant.

The total Info-Energy of an existing system is conserved in the system's multimodal Space-Time and Info-Energy structure.

3. The Law of Space-Time Conservation

The total Space-Time of an existing system equals the total Space-Time that the system possesses for the total period of its existence, including its past, present, and future. The volume of the system may change, although the total Space-Time of the system remains constant. The time of the system may change, although the total Space-Time of the system remains constant.

The total Space-Time of an existing system is conserved in the system's multimodal Space-Time and Info-Energy structure.

The Laws of Space-Time, Info-Energy, and Mass-Energy Conservation are ruled by the Laws of Space-Time, Info-Energy, and Mass-Energy Transformations, Reversibility, Symmetry, and Limitation. The Laws of Space-Time, Info-Energy, and Mass-Energy Conservation apply to every existing object and every system.

The Laws of Space-Time, Info-Energy, and Mass-Energy Transformations

1. The Law of Space-Time Transformations

Some amount of time might be transformed into the proportional amount of space, and some amount of space may be transformed into the proportional amount of time. The Space-Time transformation is a function of energy.

2. The Law of Info-Energy Transformations

Some amount of latent Info-Energy might be transformed into the proportional amount of actual Info-Energy, and some amount of actual Info-Energy may be transformed into the proportional amount of latent Info-Energy.

3. The Law of Mass-Energy Transformations

Some amount of energy might be transformed into the proportional amount of mass, and some amount of mass may be transformed into the proportional amount of energy.

The Laws of Space-Time, Info-Energy, and Mass-Energy Transformations are ruled by the Laws of Space-Time, Info-Energy, and Mass-Energy Conservation, Reversibility, Symmetry, and Limitation. The Laws of Space-Time, Info-Energy, and Mass-Energy Transformations apply to every existing object and every system.

The Laws of Space-Time, Info-Energy, and Mass-Energy Reversibility

1. The Law of Space-Time Reversibility

The Space-Time of an existing system may be reversed. The Space-Time of the system's multimodal Space-Time and Info-Energy structure may be reversed. The system's Arrows of Time and Arrows of Space may be reversed. The time and energy flow in the system's multimodal structure may be reversed.

2. The Law of Info-Energy Reversibility

The Info-Energy of an existing system may be reversed. The Info-Energy of the system's multimodal Space-Time and Info Encrgy structure may be reversed.

3.The Law of Mass-Energy Reversibility

The Mass-Energy of an existing system may be reversed. The Mass-Energy of the system's multimodal Space-Time and Info-Energy structure may be reversed. The reversibility of Space-Time, Mass-Energy, and Info-Energy is determined by the Space-Time, Info-Energy, and Mass-Energy resources of the system. If a system has reached the limits actuating a reverse (Planck length, Planck time, Planck energy or the speed of light), it is being reversed by the Matrix forces.

The Laws of Space-Time, Info-Energy, and Mass-Energy Reversibility are ruled by the Laws of Space-Time, Info-Energy, and Mass-Energy Conservation, Transformations, Symmetry, and Limitation. They apply to every existing object and every system.

Afterword

The Theory of Matrix is a new theory combining the elements of psychology, cosmology, and astrophysics. This theory was introduced by Dr Audrey E. Randles in her work 'The Theory of Matrix' in 2012.

Dr Randles developed the program for psychological investigations of the Universal Matrix along with the development of the Coresynthesis Psychological Model in early 1990s. There was a space of ten years between the first explanations of the results and association of these results with the characteristics mathematically introduced by Lorenz, Minkowski, and Einstein for the Light Cone.

When Dr Randles published her work, the Light Cone was considered a specific case applicable to the flash of light. She brought forward the idea of the universality of the Space-Time structure of the objects and a new vision on Space-Time physics.

Following the analysis of the parallels and variations between the Universal Matrix and Light Cone Dr Randles calls the Light Cone 'the Matrix of Light' and presents the Theory of Matrix as the relative importance for the true understanding of the multimodal world.

The Theory of Matrix introduces the new understanding of the multidimensional world with respect to multidimensional time. Time is no longer seen as another dimension of space, nor as a momentary feature of an event but as a multidimensional element in its own right. Dr Randles associates space with the Actuality and time with the Potentiality or Latency. Therefore, the objects are *viewed as multimodal, multidimensional objects in Space-Time.*

Books on Cosmology by Audrey E. Randles:
'Systems Theory in Cosmology' (2020)
'The Multimodal World' (2020)
'Black Holes and Supernovas' (2016)
'Grand Universe' (2016)
'Antigravity' (2015)
'Supernovas' (2015)
'The Primary Black Hole of the Universe' (2015)
'Energy in Cosmology' (2014)
'Gravity and Antigravity to the Point' (2014)
The Theory of Matrix series of books (2012 - 2013) includes the following books:
'Blocks of the Universe'
'Space and Time'
'Energy of Existence'
'Gravity and Rotation'
'Black Holes'
'Matrix of the Universe'
The New 2020 books on Cosmology include Kindle ebook and a paperback of the same title at Amazon's Book Store.

Content Use Policy

© Audrey Elizabeth Randles

Content may be used for any purpose without prior permission, subject to the special cases noted below.

By downloading the material, the user agrees:

1. to use a credit line in connection with the content. Unless otherwise noted in the caption information for any content and images the credit line should be

'Audrey E. Randles, 'The Theory of Matrix. Energy of Existence' (2013) red 2020'.

2. that we do not represent others who may claim to be authors or owners of copyright of any of the content, and make no warranties as to the quality of the content;

3. that we shall not be responsible for any loss or expenses resulting from the use of the content, and you release and hold us harmless from all liability arising from such use.

Special Cases:

This content is available for educational, journalistic, personal uses and scientific research following a scientific code of ethics.

Restrictions are placed on commercial uses. To obtain permission for commercial use, contact the copyright owner Dr Audrey E. Randles.